Simultaneous Statistical Inference

Thorsten Dickhaus

Simultaneous Statistical Inference

With Applications in the Life Sciences

Springer

Thorsten Dickhaus
Research Group "Stochastic Algorithms
and Nonparametric Statistics"
Weierstrass Institute for Applied Analysis
and Stochastics
Berlin
Germany

ISBN 978-3-642-45181-2 ISBN 978-3-642-45182-9 (eBook)
DOI 10.1007/978-3-642-45182-9
Springer Heidelberg New York Dordrecht London

Library of Congress Control Number: 2013955256

Printed on acid-free paper

Springer is part of Springer Science+Business Media (www.springer.com)

It can't be all coincidence
Too many things are evident

(Iron Maiden, Infinite Dreams, 1988)

Preface

The more questions you ask, the more wrong answers you are expected to receive—even if every single source of your information is quite trustworthy. In this work, the sources of information are data, and the questions are formalized by statistical hypothesis-alternative pairs. From the mathematical point of view, this leads to multiple test problems. We will discuss criteria and methods (in particular multiple tests) which ensure that with high probability not too many wrong decisions are made, even if many hypotheses are of interest under the scope of one and the same statistical model, i.e., regarding one and the same dataset.

High-throughput technologies in different fields of modern life sciences have led to massive multiplicity and given rise to multiple test problems with more hypotheses than observations. Driven by these developments, also new statistical paradigms have arisen. It is fair to say that a new era of multiple testing began when Yoav Benjamini and Yosef Hochberg formally introduced the false discovery rate (FDR) and the linear step-up test for FDR control in 1995. In this book, apart from classical methods controlling the family-wise error rate (FWER), theory and important life science applications of the FDR are presented in a systematic way, presumably for the first time in this depth in a monograph. In this, focus is on frequentist approaches aiming at FDR control at a fixed level. Other type I and type II error rates are mentioned and discussed where appropriate, but focus is on FWER and FDR. Chapters 6 and 7 broaden the view and show how multiple testing methodology can be used in the context of binary classification and model selection, respectively, with life science applications provided in Parts II and III. Further relationships between multiple testing and other simultaneous statistical inference problems are discussed in Chap. 1 and at respective occasions.

The book is primarily meant to be a research monograph and an introduction to simultaneous inference for applied statisticians and practitioners from the life sciences. To this end, presentation is with emphasis on applicability and we provide a couple of hints concerning which multiple test to use for which type of data. Furthermore, Chap. 8 deals with software implementing the theoretically treated procedures. However, the mainly theoretical Part I of the book may also serve as the basis for a graduate course on simultaneous statistical inference with emphasis on multiple testing for mathematical statisticians. I used parts of Chaps. 2, 4 and 5 for such a course at Humboldt-University Berlin and a couple of diploma theses in mathematics originated from this teaching.

The material for this book originated from joint work with many colleagues. I acknowledge the respective co-workers at the end of each chapter. Apart from his scientific contributions, I am especially grateful to Taras Bodnar who critically read every chapter and provided many constructive comments which helped to improve the presentation.

My deepest gratitude, however, is due to the Thai branch of my family for their enduring support in hard times. Therefore, I dedicate this work to Prayun, Duangchan, Dako, Pipat, Janyarak and her children, and my wife Supansa.

Berlin, January 2014 Thorsten Dickhaus

Contents

Part III Further Applications in the Life Sciences

Acronyms

$\mathcal{X}, \mathcal{F}, \mathcal{P}$	Statistical model
$\mathcal{X}, \mathcal{F}, \mathcal{P}, \mathcal{H}$	Multiple test problem
V_m	Number of type I errors among m tests
R_m	Total number of rejections among m tests
$\overline{\mathbb{R}}$	$\mathbb{R} \cup \{-\infty, +\infty\}$
Φ	Cumulative distribution function of the standard normal law on $\mathbb{R}$
ϕ	Lebesgue density of the standard normal law on $\mathbb{R}$
χ^2_ν	Chi-square distribution with ν degrees of freedom
t_ν	Student's t-distribution with ν degrees of freedom
$\mathrm{Beta}(a, b)$	Beta distribution with parameters a and b
$W_m(\nu, \Sigma)$	Wishart distribution with parameters m, ν and Σ
$\mathcal{M}_c(n, p)$	Multinomial distribution with c categories, sample size n and vector of probabilities p
1_A	Indicator function of the set A
$\mathcal{N}(\mu, \sigma^2)$	Normal distribution on $\mathbb{R}$ with mean μ and variance σ^2
$\mathcal{N}_k(\mu, \Sigma)$	Normal distribution on $\mathbb{R}^k$ with mean vector μ and covariance matrix Σ
$\mathcal{L}(X)$	Law (or distribution) of the random variate X
$\Gamma(\cdot)$	Gamma function, $\Gamma(x) = \int_0^\infty t^{x-1} \exp(-t) dt, x > 0$
$\det(A)$	Determinant of the matrix A
A^+	Moore-Penrose pseudo inverse of the matrix A
$\mathrm{diag}(\cdots)$	Diagonal matrix the diagonal elements of which are given by $\cdots$
$\cap$-closed	Closed under intersection
F_{ν_1, ν_2}	Fisher's F-distribution with ν_1 and ν_2 degrees of freedom
$\mathcal{B}(\mathcal{X})$	System of Borel sets of $\mathcal{X}$
$\overline{X}_{i.}$	Empirical mean in group $1 \leq i \leq k$ in k-sample problem
$\xrightarrow{w}$	Weak convergence
$\xrightarrow{d}$	Convergence in distribution
$\stackrel{d}{=}$	Equality in distribution
ABOS	Asymptotically Bayes optimal under sparsity
ANOVA	Analysis of variance
AORC	Asymptotically optimal rejection curve

BCI	Brain-computer interface
BOLD	Blood oxygen level dependent
BPI	Bonferroni plug-in
cdf	Cumulative distribution function
CRAN	Comprehensive R Archive Network
CSP	Common spatial pattern
DFM	Dynamic factor model
EC	Euler characteristic
ecdf	Empirical cumulative distribution function
EEG	Electroencephalogram
ERD	Event-related desynchronization
FCR	False coverage-statement rate
FDP	False discovery proportion
FDR	False discovery rate
fMRI	Functional magnetic resonance imaging
FWER	Family-wise error rate
GED	Generalized error distribution
GLM	Generalized linear model
HC	Higher criticism
LASSO	Least absolute shrinkage and selection operator
LD	Linkage disequilibrium
LDA	Linear discriminant analysis
LFC	Least favorable (parameter) configuration
iid.	Independent and identically distributed
MCP	Multiple comparison procedure
MLE	Maximum likelihood estimator
MSM_i	Monotonically sub-Markovian of order i
MTP	Multiple test procedure
MTP_2	Multivariate total positivity of order 2
pdf	Probability density function
pFDR	Positive false discovery rate
pFNR	Positive false non-discovery rate
PLOD	Positive lower orthant dependent
pmf	Point mass function
PRDS	Positive regression dependency on subsets
ROI	Region of interest
SD	Step-down
SNP	Single nucleotide polymorphism
SPC	Subset pivotality condition
STP	Simultaneous test procedure
SU	Step-up
SUD	Step-up-down
SVM	Support vector machine
μTOSS	Multiple hypothesis testing in an open software system
UNI$[a, b]$	Uniform distribution on the interval $[a, b]$

Chapter 1
The Problem of Simultaneous Inference

Abstract We introduce the problem of simultaneous statistical inference, with particular emphasis on testing multiple hypotheses. After a historic overview, general notation for the whole work is set up and different sources of multiplicity are distinguished. We define a variety of classical and modern type I and type II error rates in multiple hypotheses testing, analyze some relationships between them, and consider different ways to cope with structured systems of hypotheses. Relationships between multiple testing and other simultaneous statistical inference problems, in particular the construction of confidence regions for multi-dimensional parameters, as well as selection, ranking and partitioning problems, are elucidated. Finally, a general outline of the remainder of the work is given.

Simultaneous statistical inference is concerned with the problem of making several decisions simultaneously based on one and the same dataset. In this work, simultaneous statistical decision problems will mainly be formalized by multiple hypotheses and multiple tests. Not all simultaneous statistical decision problems are given in this formulation in the first place, but they can often be re-formulated in terms of multiple test problems. General relationships between multiple testing and other kinds of simultaneous statistical decision problems will briefly be discussed in Sect. 1.3. Moreover, we will refer to specific connections at respective occasions. For instance, we will elucidate connections between multiple testing and binary classification in Chap. 6 and discuss multiple testing methods in the context of model selection in Chap. 7.

The origins of multiple hypotheses testing can at least be traced back to Bonferroni (1935, 1936). The "Bonferroni correction"(cf. Example 3.1) is a generic method for evaluating several statistical tests simultaneously and ensuring that the probability for *at least one* type I error is bounded by a pre-defined significance level α. The latter criterion is nowadays referred to as (strong) control of the family-wise error rate (FWER) at level α and will be defined formally in Definition 1.2 below. In well-defined model classes, the Bonferroni method can be improved. In the 1950s, especially analysis of variance (ANOVA) models have been studied with respect to multiple comparisons of group-specific means. For instance, Tukey (1953)

T. Dickhaus, *Simultaneous Statistical Inference*,
DOI: 10.1007/978-3-642-45182-9_1,

developed a multiple test for all pairwise comparisons of means in ANOVA models based on the studentized range distribution. Keuls (1952) applied this technique to a ranking problem of ANOVA means in an agricultural context. The works of Dunnett (1955, 1964) treated the problem of multiple comparisons with a control group, while Scheffé (1953) provided a method for testing general linear contrasts simultaneously in the ANOVA context.Concepts from multivariate analysis and probability theory, in particular multivariate dependency concepts, have also been used for multiple testing, cf. for instance the works by Šidák (1967, 1968, 1971, 1973). These concepts allow for establishing probability bounds which in turn can be used for adjusting significance levels for multiplicity. We will provide details in Sect. 4.3. While all the aforementioned historical methods lead to single-step tests (meaning that the same, multiplicity-adjusted critical value is used for all test statistics corresponding to the considered tests), the formal introduction of the closed test principle by Marcus et al. (1976) paved the way for stepwise rejective multiple tests (for a detailed description of these different classes of multiple test procedures, see Chap. 3). These stepwise rejective tests are often improvements of the classical single-step tests with respect to power, meaning that they allow (on average) for more rejections of false hypotheses while controlling the same type I error criterion (namely, the FWER at a given level of significance). Stepwise rejective FWER-controlling multiple tests have been developed in the late 1970s, the 1980s and early 1990s; see, for example, Holm (1977, 1979), Hommel (1988) (based on Simes (1986)), Hochberg (1988), and Rom (1990). Around this time, the theory of FWER control had reached a high level of sophistication and was treated in the monographs by Hochberg and Tamhane (1987) and Hsu (1996).

It is fair to say that a new era of multiple testing began when Benjamini and Hochberg (1995) introduced a new type I error criterion, namely control of the false discovery rate (FDR), see Definition 1.2. Instead of bounding the probability of one or more type I errors, the FDR criterion bounds the expected proportion of false positives among all significant findings, which typically implies to allow for a few type I errors; see also Seeger (1968) and Sorić (1989) for earlier instances of this idea. During the past 20 years, simultaneous statistical inference and, in particular, multiple statistical hypothesis testing has become a major branch of mathematical and applied statistics, cf. Benjamini (2010) for some bibliometric details. Even for experts it is hardly possible to keep track of the exponentially (over time) growing literature in the field. This growing importance is not least due to the data-analytic challenges posed by large-scale experiments in modern life sciences such as, for instance, genetic association studies (cf. Chap. 9), gene expression studies (Chap. 10), functional magnetic resonance imaging (Chap. 11), and brain-computer interfacing (Chap. 12). Hence, the present work is attempting to explain some of the most important theoretical basics of simultaneous statistical inference, together with applications in diverse areas of the life sciences.

1.1 Sources of Multiplicity

The following definition is fundamental for the remainder of this work.

Definition 1.1 (Statistical model). A statistical model is a triple $(\mathscr{X}, \mathscr{F}, \mathscr{P})$. In this, $\mathscr{X}$ denotes the sample space (the set of all possible observations), $\mathscr{F}$ a σ-field on $\mathscr{X}$ (the set of all events that we can assign a probability to) and $\mathscr{P}$ a family of probability measures on the measurable space $(\mathscr{X}, \mathscr{F})$. Often, we will write $\mathscr{P}$ in the form $(\mathbb{P}_\vartheta)_{\vartheta \in \Theta}$, such that the family is indexed by the parameter ϑ of the model which can take values in the parameter space Θ, where Θ may have infinite dimension. Unless stated otherwise, an observation will be denoted by $x \in \mathscr{X}$, and we think of x as the realization of a random variate X which mathematically formalizes the data-generating mechanism. The target of statistical inference is the parameter ϑ which we regard as the unknown and unobservable state of nature.

Once the statistical model for the data-generating process at hand is defined, two general types of resulting multiplicity can be labeled as "one- or two- sample problems with multiple endpoints" and "k-sample problems with localized comparisons", where $k > 2$, respectively. In one- or two- sample problems with multiple endpoints, the sample space is often of the form $\mathscr{X} = \mathbb{R}^{m \times n}$. The same n observational units are measured with respect to m different endpoints, where we assumed for ease of presentation that every measurement results in a real number. The transfer to measurements of other type (for instance, allele pairs at genetic loci) is straightforward. For every of the m endpoints, an own scientific question can be of interest. On the contrary, in k-sample problems with localized comparisons, the sample space is typically of the form $\mathscr{X} = \mathbb{R}^{\sum_{i=1}^k n_i}$, meaning that $k > 2$ different groups of observational units (for instance, corresponding to k different doses of a drug) are considered, and that n_i observations are made in group i, where $1 \leq i \leq k$. In this, all $\sum_{i=1}^k n_i$ measurements concern one and the same endpoint (for instance, a disease status). The scientific questions in the latter case typically relate to differences between the k groups. Multiplicity arises, if not (only) general homogeneity or heterogeneity between the groups shall be assessed, but if differences, if any, are to be localized in the sense that we want to find out *which* groups are different. Two classical examples are the "all pairs" problem (all $m = k(k-1)/2$ pairwise group comparisons are of interest) and the "multiple comparisons with a control" problem (group k is a reference group and all other $m = k - 1$ groups are to be compared with group k).

We will primarily focus on these two kinds of problems. However, it has to be mentioned that they do not cover the whole spectrum of simultaneous statistical inference problems. For instance, flexible (group-sequential and adaptive) study designs induce a different type of multiplicity problem that we will not consider in the present work.

Throughout the remainder, we will try to stick to the notation developed in this section: m is the number of comparisons (the multiplicity of the problem), n or a subscripted n denotes a sample size and k refers to the total number of groups in a

k-sample problem or to the dimensionality of the parameter ϑ. Often, the two latter quantities are identical.

1.2 Multiple Hypotheses Testing

In what follows, we (sometimes implicitly) identify statistical hypotheses with non-empty subsets of the parameter space Θ. The tuple $(\mathcal{X}, \mathcal{F}, (\mathbb{P}_\vartheta)_{\vartheta \in \Theta}, \mathcal{H})$ denotes a multiple test problem, where $\mathcal{H} = (H_i : i \in I)$ for an arbitrary index set I defines a family of null hypotheses. The resulting alternative hypotheses are denoted by $K_i = \Theta \setminus H_i, i \in I$. The intersection hypothesis $H_0 = \bigcap_{i \in I} H_i$ will be referred to as global hypothesis. Throughout the work, we assume that H_0 is non-empty.With very few exceptions, we will consider the case of finite families of hypotheses, meaning that $|I| = m \in \mathbb{N}$. In such cases, we will often write $\mathcal{H}_m$ instead of $\mathcal{H}$ and index the hypotheses such that $I = \{1, \ldots, m\}$. A (non-randomized) multiple test for $(\mathcal{X}, \mathcal{F}, (\mathbb{P}_\vartheta)_{\vartheta \in \Theta}, \mathcal{H}_m)$ is a measurable mapping $\varphi = (\varphi_i)_{1 \leq i \leq m} : \mathcal{X} \to \{0, 1\}^m$ the components of which have the usual interpretation of a statistical test for H_i versus K_i. Namely, H_i is rejected if and only if $\varphi_i(x) = 1$, where $x \in \mathcal{X}$ denotes the observed data.

1.2.1 Measuring and Controlling Errors

The general decision pattern of a multiple test for m hypotheses is summarized in Table 1.1. In contrast to usual, one-dimensional test problems, it becomes apparent that type I and type II errors can occur simultaneously. In Table 1.1, type I errors are counted by V_m and type II errors are counted by T_m. The total number of rejections is denoted by R_m. Notice that the quantities U_m, V_m, T_m, S_m and m_0, m_1 all depend on the unknown value of the parameter ϑ (although we suppressed this dependence on ϑ notationally in Table 1.1) and are therefore unobservable. Only m and R_m can be observed.

For a given $\vartheta \in \Theta$, we denote the index set of true null hypotheses in $\mathcal{H}_m$ by $I_0 \equiv I_0(\vartheta) = \{1 \leq i \leq m : \vartheta \in H_i\}$. Analogously, we define $I_1 \equiv I_1(\vartheta) = I \setminus I_0$. With this notation, we can formally define $V_m \equiv V_m(\vartheta) = |\{i \in I_0(\vartheta) : \varphi_i = 1\}|$, $S_m \equiv S_m(\vartheta) = |\{i \in I_1(\vartheta) : \varphi_i = 1\}|$, and $R_m \equiv R_m(\vartheta) = |\{i \in I : \varphi_i =$

Table 1.1 Decision pattern of a multiple test procedure

	Test decisions		
Hypotheses	0	1	
True	U_m	V_m	m_0
False	T_m	S_m	m_1
	W_m	R_m	m

$1\}| = V_m + S_m$. Based on these quantities, the following definition is concerned with measuring and controlling type I errors of a multiple test φ.

Definition 1.2 (Multiple type I error rates). Let $(\mathcal{X}, \mathcal{F}, (\mathbb{P}_\vartheta)_{\vartheta \in \Theta}, \mathcal{H}_m)$ denote a multiple test problem and $\varphi = (\varphi_i : i \in I)$ a multiple test for $(\mathcal{X}, \mathcal{F}, (\mathbb{P}_\vartheta)_{\vartheta \in \Theta}, \mathcal{H}_m)$.

(a) The number

$$\mathrm{FWER}_\vartheta(\varphi) = \mathbb{P}_\vartheta(V_m > 0) = \mathbb{P}_\vartheta\left(\bigcup_{i \in I_0(\vartheta)} \{\varphi_i = 1\}\right)$$

is called the family-wise error rate (FWER) of φ under ϑ.

(b) The random variable

$$\mathrm{FDP}_\vartheta(\varphi) = \frac{V_m}{\max(R_m, 1)}$$

is called the false discovery proportion (FDP) of φ under ϑ.

(c) The number

$$\mathrm{FDR}_\vartheta(\varphi) = \mathbb{E}_\vartheta[\mathrm{FDP}_\vartheta(\varphi)] = \mathbb{E}_\vartheta\left[\frac{V_m}{\max(R_m, 1)}\right]$$

is called the false discovery rate (FDR) of φ under ϑ.

(d) The number

$$\mathrm{pFDR}_\vartheta(\varphi) = \mathbb{E}_\vartheta\left[\frac{V_m}{R_m} \mid R_m > 0\right]$$

is called the positive false discovery rate (pFDR) of φ under ϑ.

(e) The multiple test φ is called a multiple test at local level $\alpha \in (0, 1)$, if each φ_i is a level α test for H_i versus K_i.

(f) The multiple test φ is said to control the FWER in the strong sense (strongly) at level $\alpha \in (0, 1)$, if

$$\sup_{\vartheta \in \Theta} \mathrm{FWER}_\vartheta(\varphi) \leq \alpha. \tag{1.1}$$

(g) The multiple test φ is said to control the FWER in the weak sense (weakly) at level $\alpha \in (0, 1)$, if

$$\forall \vartheta \in H_0 : \mathrm{FWER}_\vartheta(\varphi) \leq \alpha. \tag{1.2}$$

(h) The multiple test φ is said to control the FDR at level $\alpha \in (0, 1)$, if

$$\sup_{\vartheta \in \Theta} \mathrm{FDR}_\vartheta(\varphi) \leq \alpha. \tag{1.3}$$

(i) We call a parameter value ϑ^* a least favourable parameter configuration (LFC) for the FWER or the FDR, respectively, of a given multiple test φ, if ϑ^* yields the supremum in (1.1) or (1.3), respectively.

The following lemma, though obvious, will be useful for the construction of closed test procedures, see Sect. 3.3.

Lemma 1.1. *Let $(\mathscr{X}, \mathscr{F}, (\mathbb{P}_\vartheta)_{\vartheta \in \Theta}, \mathscr{H})$ denote a multiple test problem and $\varphi = (\varphi_i : i \in I)$ a multiple test for $(\mathscr{X}, \mathscr{F}, (\mathbb{P}_\vartheta)_{\vartheta \in \Theta}, \mathscr{H})$.*

(a) Strong FWER control of φ implies weak FWER control of φ.
(b) Assume that φ controls the FWER weakly at level α. Then, a level α test for the (single) global hypothesis H_0 is given by the following rule: Reject H_0 if there exists an $i \in I$ such that $\varphi_i(x) = 1$.

For the relationships between the FWER, the FDR, and the pFDR, the following assertions hold true.

Lemma 1.2 (Relationships between FWER, FDR and pFDR). *Under the assumptions of Definition 1.2, we get:*

(a) $FDR_\vartheta(\varphi) = pFDR_\vartheta(\varphi)\mathbb{P}_\vartheta(R_m > 0)$.
(b) If $m_0(\vartheta) = m$, then $FDR_\vartheta(\varphi) = FWER_\vartheta(\varphi)$.
(c) For any $\vartheta \in \Theta$, it holds $FDR_\vartheta(\varphi) \leq FWER_\vartheta(\varphi)$.

Proof. To prove part (a), we calculate straightforwardly

$$\begin{aligned}
\mathrm{FDR}_\vartheta(\varphi) &= \mathbb{E}_\vartheta\left[\frac{V_m}{R_m \vee 1}\right] \\
&= \mathbb{E}_\vartheta\left[\frac{V_m}{R_m \vee 1} | R_m > 0\right]\mathbb{P}_\vartheta\,(R_m > 0) \\
&\quad + \mathbb{E}_\vartheta\left[\frac{V_m}{R_m \vee 1} | R_m = 0\right]\mathbb{P}_\vartheta\,(R_m = 0) \\
&= \mathrm{pFDR}_\vartheta(\varphi)\mathbb{P}_\vartheta\,(R_m > 0) + 0.
\end{aligned}$$

For the proof of part (b), we notice that, if $m_0 = m$, $V_m = R_m$. Hence, $\mathrm{pFDR}_\vartheta(\varphi) \equiv 1$ in this case and, making use of part (a),

$$\mathrm{FDR}_\vartheta(\varphi) = \mathbb{P}_\vartheta\,(R_m > 0) = \mathbb{P}_\vartheta\,(V_m > 0) = \mathrm{FWER}_\vartheta(\varphi).$$

In the general case, we easily verify that $\mathrm{FDP}_\vartheta(\varphi) \leq \mathbf{1}_{\{V_m > 0\}}$. Thus, $\mathbb{E}_\vartheta\,[\mathrm{FDP}_\vartheta(\varphi)] \leq \mathbb{E}_\vartheta\left[\mathbf{1}_{\{V_m > 0\}}\right]$, which is equivalent to the assertion of part (c). □

Notice that the proof of part (b) of Lemma 1.2 implies that the pFDR cannot be controlled in the frequentist sense. The pFDR is only useful in Bayesian considerations (cf., e.g., Chap. 6). Throughout the remainder of this work, we will restrict our attention to the type I error rates defined in Definition 1.2. This is mainly due to the fact that they are most commonly applied in practice. However, let us mention a few additional type I error rates that are occasionally found in multiple testing literature. While the terms in Definition 1.2 are quite standard, the following quantities have been introduced under a variety of different names and acronyms by different authors.

Definition 1.3 (Further type I error rates). Under the assumptions of Definition 1.2, the following quantities are alternative type I error rates in multiple hypotheses testing.

(i) For a fixed positive integer k, the number

$$k\text{-FWER}_{\vartheta}(\varphi) = \mathbb{P}_{\vartheta}(V_m > k)$$

is called the generalized family-wise error rate (k-FWER) of φ under ϑ.

(ii) The number

$$\text{ENFR}_{\vartheta}(\varphi) = \mathbb{E}_{\vartheta}[V_m]$$

is called the expected number of false rejections (ENFR) of φ under ϑ.

(iii) The number

$$\text{EER}_{\vartheta}(\varphi) = \mathbb{E}_{\vartheta}[V_m/m] = \text{ENFR}_{\vartheta}(\varphi)/m$$

is called the expected (type I) error rate (EER) of φ under ϑ.

(iv) For a given constant $c \in (0, 1)$, the number

$$\text{FDX}_{\vartheta}(\varphi) = \mathbb{P}_{\vartheta}(\text{FDP}_{\vartheta}(\varphi) > c)$$

is called the false discovery exceedance rate (FDX) of φ under ϑ.

In order to compare concurring multiple test procedures (which should control the same type I error rate at the same level), also a type II error measure or, equivalently, a notion of power is required under the multiple testing framework. The most popular notion of multiple power is defined as follows.

Definition 1.4. Under the assumptions of Definition 1.2, we call

$$\text{power}_{\vartheta}(\varphi) = \mathbb{E}_{\vartheta}[S_m/\max(m_1, 1)] \tag{1.4}$$

the multiple power of φ under ϑ.

We may remark here that our Definition 1.4 is in conflict with the nomenclature of Maurer and Mellein (1988) who referred to the right-hand side of (1.4) as the expected average power of φ.

1.2.2 Structured Systems of Hypotheses

Identifying hypotheses with subsets of the parameter space allows us to apply set-theoretic operations to them. In particular, we can analyze subset/superset relations among the elements in $\mathscr{H}_m$.

Definition 1.5 (Structured systems of hypotheses). Let $\mathscr{H}_m = \{H_i : i \in I = \{1, \ldots, m\}\}$ denote a finite system of hypotheses.

(a) A hypothesis $H_i \in \mathscr{H}_m$ is called an elementary hypothesis, if H_i cannot be written as an intersection of strict supersets of H_i in $\mathscr{H}_m$.
(b) A hypothesis $H_i \in \mathscr{H}_m$ is called a minimal hypothesis, if H_i does not possess strict supersets in $\mathscr{H}_m$.
(c) A hypothesis $H_i \in \mathscr{H}_m$ is called a maximal hypothesis, if $\mathscr{H}_m$ contains no element of which H_i is a strict superset.
(d) The system $\mathscr{H}_m$ is called closed under intersection ($\cap$-closed), if $\forall \emptyset \neq J \subseteq I$: $H_J = \bigcap_{j \in J} H_j = \emptyset$ or $H_J \in \mathscr{H}_m$.
(e) The system $\mathscr{H}_m$ is called hierarchically structured (hierarchical for short) if at least one element of $\mathscr{H}_m$ has a strict superset in $\mathscr{H}_m$.

Throughout the remainder, we will graphically illustrate hierarchies in structured systems of hypotheses in the following two ways.

Example 1.1. Let $m = 3$ and $\mathscr{H} = \{H_1, H_2, H_3\}$.

(a) A very useful way to illustrate subset/superset relations in $\mathscr{H}$ is an arrow diagram.

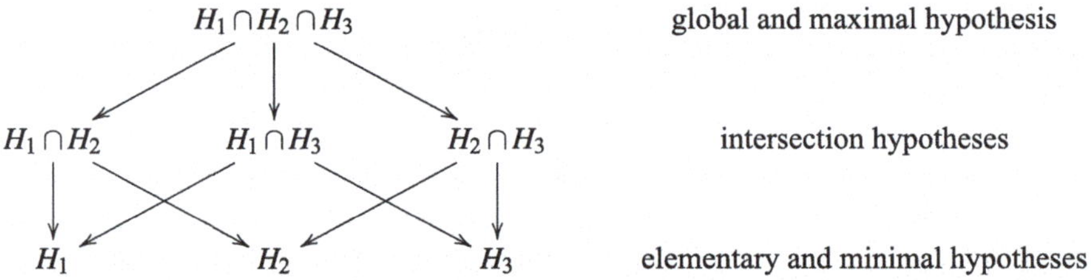

The arrows point to the hypotheses which are the corresponding supersets (implications).

(b) Alternatively, one may also draw a Venn diagram.

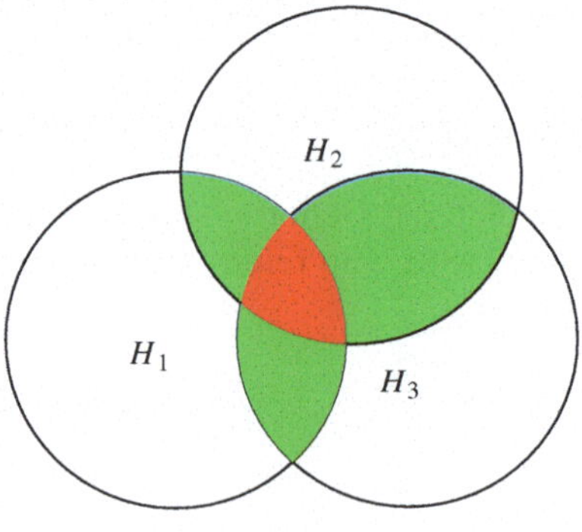

If logical restrictions (subset/superset relations) exist in $\mathcal{H}_m$, then a suitable multiple test procedure for $(\mathcal{X}, \mathcal{F}, (\mathbb{P}_\vartheta)_{\vartheta \in \Theta}, \mathcal{H}_m)$ should not lead to decision patterns which contradict this logical structure. Two important properties are coherence and consonance.

Definition 1.6 (Gabriel (1969)). Let φ denote a multiple test for $(\mathcal{X}, \mathcal{F}, \mathcal{P}, \mathcal{H}_m)$.

(a) The test φ is called coherent, if

$$\forall i, j \in I \quad \text{with} \quad H_i \subseteq H_j : \{\varphi_j = 1\} \subseteq \{\varphi_i = 1\}.$$

Thus, if a coherent multiple test rejects a particular hypothesis H_j, then it necessarily also rejects all subsets of H_j. Otherwise, φ is called incoherent.

(b) The test φ is called consonant, if

$$\forall i \in I \quad \text{with} \quad \exists j \in I : H_i \subset H_j : \{\varphi_i = 1\} \subseteq \bigcup_{j : H_j \supset H_i} \{\varphi_j = 1\}.$$

Thus, if a particular hypothesis H_i is rejected by a consonant multiple test and it possesses strict supersets in $\mathcal{H}_m$, then necessarily at least one of these strict supersets is also rejected. Otherwise, φ is called dissonant.

In practice, coherence is an indispensable requirement which any multiple test should fulfill. Consonance is generally desirable, too, but enforcing it can lead to very conservative procedures (with very few rejections, if any). In Sect. 3.3, we will introduce a general construction principle for coherent FWER-controlling multiple test procedures, namely, the closure principle.However, closed test procedures are in general not consonant.

1.3 Relationships to Other Simultaneous Statistical Inference Problems

Multiple testing methodology is not only useful in itself (i.e., for actually testing multiple hypotheses), but also for solving other, related simultaneous statistical decision problems. The Habilitationsschrift of Finner (1994) provides an in-depth analysis of connections between multiple testing and such other problems.

Maybe, the most straightforward connection can be drawn to the problem of constructing (simultaneous) confidence regionsfor multi-dimensional parameters. The general solution to this problem by multiple testing methodology is given in Theorem 1.1.

Theorem 1.1 (Extended Correspondence Theorem, see Sect. 4.1 in Finner (1994)). *Let $\mathcal{H} = \{H_i : i \in I\}$ denote an arbitrary family of hypotheses and $(\mathcal{X}, \mathcal{F}, \mathcal{P}, \mathcal{H})$ a multiple test problem.*

(a) *If* $\varphi = (\varphi_i : i \in I)$ *is a strongly FWER-controlling multiple test at FWER level* α *for* $(\mathscr{X}, \mathscr{F}, \mathscr{P}, \mathscr{H})$ *and we define* $C(x) = \bigcap_{j:\varphi_j(x)=1} K_j$, $x \in \mathscr{X}$, *with the convention* $\bigcap_{j\in\emptyset} K_j = \Theta$, *then* $\mathscr{C} = \mathscr{C}(\varphi) = (C(x) : x \in \mathscr{X})$ *constitutes a family of confidence regions for* $\vartheta \in \Theta$ *at confidence level* $1-\alpha$.

(b) *Assume that a family* $\mathscr{C} = (C(x) : x \in \mathscr{X})$ *of confidence regions at confidence level* $1-\alpha$ *for* $\vartheta \in \Theta$ *is given. Define the multiple test* φ *by* $\varphi(\mathscr{C}) = (\varphi_i : i \in I)$, *where* $\varphi_i(x) = \mathbf{1}_{K_i}(C(x))$ *for all* $x \in \mathscr{X}$ *and all* $i \in I$. *Then,* $\varphi(\mathscr{C})$ *is a strongly FWER-controlling multiple test at FWER level* α *for* $(\mathscr{X}, \mathscr{F}, \mathscr{P}, \mathscr{H})$.

For instance, part (b) of Theorem 1.1 is the basis for Scheffé tests, see Theorem 3.1. Further connections exist to selection, partitioning and ranking problems, cf., for instance, Gupta and Panchapakesan (1979), Liese and Miescke (2008) and references within. Let us briefly discuss one example of a selection problem.

Example 1.2. Consider the model of the one-factorial analysis of variance with balanced design, meaning that the data-generating mechanism can be represented by a random matrix $X = (X_{ij})$, where $1 \leq i \leq k$ and $1 \leq j \leq n$, with k denoting the number of groups and n the common sample size per group. Assume that all X_{ij} are jointly stochastically independent, real-valued random variables with $X_{ij} \sim \mathscr{N}(\mu_i, \sigma^2)$. In this, $\mu_i \in \mathbb{R}$ for $1 \leq i \leq k$ denotes the group-specific mean and $\sigma^2 > 0$ the common (known or unknown) variance. Assume that $k \geq 3$ and $n \geq 2$. In the notation of a statistical model, we consequently have $\mathscr{X} = \mathbb{R}^{k \cdot n}$, $\mathscr{F} = \mathbb{B}^{k \cdot n}$ (the Borel σ-field on $\mathbb{R}^{k \cdot n}$), and $\Theta = \mathbb{R}^k \times [0, \infty)$ with parameter $\vartheta = (\mu_1, \ldots, \mu_k, \sigma^2)^\top$. For ease of argumentation, assume that the k groups correspond to k different treatments and that μ_i quantifies the i-th mean treatment effect. Let the aim of the statistical analysis be to find (select) the best treatment(s) (with largest μ_i). For example, we may define the set of good treatments as

$$G(\vartheta) = \{i : \max_{1 \leq j \leq k} \mu_j - \mu_i \leq \varepsilon\sigma\}$$

for some given constant $\varepsilon \geq 0$; see equation (1.3) in Finner and Giani (1994), for example. Bechhofer (1954) studied the case of $\varepsilon = 0$ and known σ^2. Letting $Y_i = \bar{X}_{i.}$, $1 \leq i \leq k$, denote the empirical group means and $Y_{1:k} \leq \cdots \leq Y_{k:k}$ their order statistics, it is near at hand to select the treatment corresponding to $Y_{k:k}$. The question arises if one can guarantee (for instance, by choosing n large enough) that this decision rule selects the actually best treatment with a probability at least equal to a given bound P^* (say). The answer to this question is in general negative, as can be seen by multiple testing considerations. Let $\mu_{[i]} \leq \cdots \leq \mu_{[k]}$ denote the ordered theoretical means and $Y_{(i)} \sim \mathscr{N}(\mu_{[i]}, \sigma^2/n)$ for $1 \leq i \leq k$ the correspondingly re-arranged empirical means. Furthermore, let Z_i, $1 \leq i \leq k$, denote iid. standard normal random variables. Then we can explicitly calculate the probability of a correct selection (PCS) by

$$
\begin{aligned}
&\mathbb{P}_{\vartheta}(\forall 1 \leq i \leq k-1 : Y_{(k)} \geq Y_{(i)}) \\
&= \mathbb{P}_{(\mathbf{0},1)}\left(\forall 1 \leq i \leq k-1 : \frac{\sigma}{\sqrt{n}} Z_k + \mu_{[k]} \geq \frac{\sigma}{\sqrt{n}} Z_i + \mu_{[i]}\right) \\
&= \int_{-\infty}^{\infty} \mathbb{P}_{(\mathbf{0},1)}\left(\forall 1 \leq i \leq k-1 : Z_i \leq Z_k + \frac{\sqrt{n}}{\sigma}(\mu_{[k]} - \mu_{[i]}) | Z_k = z\right) d\mathbb{P}^{Z_k}(z) \\
&= \int_{-\infty}^{\infty} \prod_{i=1}^{k-1} \Phi\left(z + \frac{\sqrt{n}}{\sigma}(\mu_{[k]} - \mu_{[i]})\right) \phi(z) dz,
\end{aligned} \tag{1.5}
$$

where Φ and ϕ denote the cdf and the pdf of the standard normal distribution on $\mathbb{R}$, respectively. It is easy to check that (1.5) is equal to $1/k$ independently of $\sigma^2 > 0$ and $n \geq 2$ if $\mu_1 = \cdots = \mu_k$. Hence, in general it is only possible to keep a PCS level of $1/k$ which is clearly unsatisfactory. Thus, Bechhofer (1954) restricted his attention to a parameter subspace $\Theta^* = \{\vartheta : \mu_{[k]} - \mu_{[k-1]} \geq \varepsilon\sigma\}$ for a fixed constant $\varepsilon \equiv \varepsilon(n) > 0$, the so-called "preference zone".

1.4 Contributions of this Work

After this introductory chapter, we divide the material into three parts. The first part contributes to mathematical statistics and contains general methodological ideas. We first discuss the concept of p-values which is important for multiple hypotheses testing (Chap. 2). In Chap. 3 we attempt to provide a general overview in terms of a systematization of multiple test procedures with respect to error control, structure of the decision rule and degree of detail of the underlying statistical model. Then, we investigate specific classes of multiple tests in more detail, namely, simultaneous test procedures (Chap. 4) and stepwise rejective multiple tests (Chap. 5). To provide some applications in and draw connections to other areas of statistics, we describe relationships between multiple testing and binary classification (Chap. 6) and model selection (Chap. 7), respectively. We conclude Part I with some comments on software solutions for multiple hypotheses testing (Chap. 8) for a smooth transition to the following parts.

Parts II and III are then devoted to practical applications of multiple test procedures in the life sciences. Part II considers statistical genetics and, in particular, the problem of detecting associations between a binary phenotype and genetic profiles in humans. We describe three stages of decreasing biological distance between genotype and phenotype: (i) association analysis based on genetic markers (Chap. 9), (ii) gene expression analysis (Chap. 10), (iii) functional confirmation (in particular functional magnetic resonance imaging, see Chap. 11). In all three stages, multiple testing methodology is a helpful tool, but the specific characteristics of the stages require different fine-tuning of multiple tests for the respective purposes. In Part III, we investigate several other application areas from the life sciences. This collection is neither meant to be exhaustive nor representative, but is merely due to the author's experience in applications.

This work is addressed to mathematical statisticians and practitioners from the life sciences. Therefore, we will precisely state and, where appropriate, prove the general results in Part I. Nevertheless, we also explain all main results and techniques in verbal form, such that practitioners who do not want to study the mathematics in all detail can follow the exposition. However, we have to assume that the reader has basic knowledge in statistical test theory. In particular, she should be familiar with the concepts of type I and type II errors, significance level, power, Neyman-Pearson tests and likelihood ratio tests.

Acknowledgments Some parts of this chapter were inspired by material from unpublished lecture notes by Helmut Finner and Iris Pigeot. I thank Klaus Straßburger for many fruitful discussions, especially about simultaneous confidence regions. Special thanks are due to Mareile Große Ruse for programming parts of the LaTeX code used in this chapter.

References

Bechhofer R (1954) A single-sample multiple decision procedure for ranking means of normal populations with known variances. Ann Math Stat 25:16–39. doi:10.1214/aoms/1177728845

Benjamini Y (2010) Simultaneous and selective inference: current successes and future challenges. Biom J 52(6):708–721

Benjamini Y, Hochberg Y (1995) Controlling the false discovery rate: a practical and powerful approach to multiple testing. J R Stat Soc Ser B Stat Methodol 57(1):289–300

Bonferroni CE (1935) Il calcolo delle assicurazioni su gruppi di teste. Studi in onore Salvatore Ortu Carboni 13–60

Bonferroni CE (1936) Teoria statistica delle classi e calcolo delle probabilita. Pubbl. d. R. Ist. Super. di Sci. Econom. e Commerciali di Firenze 8. Firenze: Libr. Internaz. Seeber

Dunnett CW (1955) A multiple comparison procedure for comparing several treatments with a control. J Am Stat Assoc 50:1096–1121. doi:10.2307/2281208

Dunnett CW (1964) New tables for multiple comparisons with a control. Biometrics 20:482–491

Finner H (1994) Testing multiple hypotheses: general theory, specific problems, and relationships to other multiple decision procedures. Fachbereich IV, Universität Trier, Habilitationsschrift

Finner H, Giani G (1994) Closed subset selection procedures for selecting good populations. J Stat Plann Infer 38(2):179–199. doi:10.1016/0378-3758(94)90034-5

Gabriel KR (1969) Simultaneous test procedures—some theory of multiple comparisons. Ann Math Stat 40:224–250. doi:10.1214/aoms/1177697819

Gupta SS, Panchapakesan S (1979) Multiple decision procedures: theory and methodology of selecting and ranking populations. Wiley Series in Probability and Mathematical Statistics. A Wiley Publication in Applied Statistics. Wiley, New York

Hochberg Y (1988) A sharper Bonferroni procedure for multiple tests of significance. Biometrika 75(4):800–802. doi:10.1093/biomet/75.4.800

Hochberg Y, Tamhane AC (1987) Multiple comparison procedures. Wiley Series in Probability and Mathematical Statistics. Applied Probability and Statistics. Wiley, New York

Holm SA (1977) Sequentially rejective multiple test procedures. Statistical Research Report No. 1977-1. Institute of Mathematics and Statistics, University of Umeå.

Holm SA (1979) A simple sequentially rejective multiple test procedure. Scand J Stat Theory Appl 6:65–70

Hommel G (1988) A stagewise rejective multiple test procedure based on a modified Bonferroni test. Biometrika 75(2):383–386. doi:10.1093/biomet/75.2.383
Hsu JC (1996) Multiple comparisons: theory and methods. Chapman and Hall, London
Keuls M (1952) The use of the "Studentized Range" in connection with an analysis of variance. Euphytica 1:112–122
Liese F, Miescke KJ (2008) Statistical decision theory. Estimation, testing, and selection. Springer Series in Statistics. Springer, New York, doi:10.1007/978-0-387-73194-0
Marcus R, Peritz E, Gabriel KR (1976) On closed test procedures with special reference to ordered analysis of variance. Biometrika 63(3):655–660
Maurer W, Mellein B (1988) On new multiple tests based on independent p-values and the assessment of their power. In: Bauer P, Hommel G, Sonnemann E (eds) Multiple hypothesenprüfung—multiple hypotheses testing. Symposium Gerolstein 1987, Springer, Berlin. Medizinische Informatik und Statistik 70, pp 48–66
Rom DM (1990) A sequentially rejective test procedure based on a modified Bonferroni inequality. Biometrika 77:663–665
Scheffé H (1953) A method for judging all contrasts in the analysis of variance. Biometrika 40:87–110
Seeger P (1968) A note on a method for the analysis of significances en masse. Technometrics 10(3):586–593
Šidák Z (1967) Rectangular confidence regions for the means of multivariate normal distributions. J Am Stat Assoc 62:626–633. doi:10.2307/2283989
Šidák Z (1968) On multivariate normal probabilities of rectangles: their dependence on correlations. Ann Math Stat 39:1425–1434
Šidák Z (1971) On probabilities of rectangles in multivariate student distributions: their dependence on correlations. Ann Math Stat 42:169–175. doi:10.1214/aoms/1177693504
Šidák Z (1973) On probabilities in certain multivariate distributions: their dependence on correlations. Apl Mat 18:128–135
Simes RJ (1986) An improved Bonferroni procedure for multiple tests of significance. Biometrika 73:751–754
Sorić B (1989) Statistical "Discoveries" and effect-size estimation. J Am Stat Assoc 84(406):608–610
Tukey JW (1953) The problem of multiple comparisons. In: Braun HI, Kaplan B, Sheehan KM, Wang M-H (eds) The collected works of John W. Tukey. Volume VIII: multiple comparisons: 1948–1983. Chapman and Hall, New York, pp 1–300

Part I
General Theory

Chapter 2
Some Theory of p-values

Abstract Many multiple test procedures are formalized and carried out in practice by means of p-values. In this chapter, we formally introduce the notion of a p-value and its usage for testing a statistical hypothesis. Methods for computing p-values are discussed with respect to tests of Neyman-Pearson type and for discrete statistical models. In the context of testing multiple hypotheses, we introduce the concept of local significance levels. Randomized p-values are discussed for situations with multiple composite hypotheses and for discretely distributed test statistics. Some p-value models commonly used in multiple testing literature are explained. In view of stepwise rejective multiple test procedures, properties of order statistics of p-values are discussed for some of these models.

Many (stepwise) multiple tests are formalized and carried out by means of p-values corresponding to (marginal) test statistics. In the statistical literature, there exists an overwhelming debate whether p-values are suitable decision tools, cf. the references in Sect. 3.11 of Lehmann and Romano (2005). In this work, we pragmatically regard a p-value as a deterministic transformation of a test statistic which is particularly useful for multiple testing, because it provides a standardization. Every p-value is supported on the unit interval [0, 1], even if test statistics have drastically different scales.

Definition 2.1 (***p*-value**). Let $(\mathscr{X}, \mathscr{F}, (\mathbb{P}_\vartheta)_{\vartheta \in \Theta})$ a statistical model and φ a (one-dimensional) non-randomized test for the single pair of hypotheses $\emptyset \neq H \subset \Theta$ versus $K = \Theta \setminus H$. Assume that φ is based on a real-valued test statistic $T : \mathscr{X} \to \mathbb{R}$. More specifically, let φ be characterized by rejection regions $\Gamma_\alpha \subset \mathbb{R}$ for any given significance level $\alpha \in (0, 1)$, such that $\varphi(x) = 1 \iff T(x) \in \Gamma_\alpha$ for $x \in \mathscr{X}$. Then, we define the p-value of an observation $x \in \mathscr{X}$ with respect to φ by

$$p_\varphi(x) = \inf_{\{\alpha : T(x) \in \Gamma_\alpha\}} \mathbb{P}^*(T(X) \in \Gamma_\alpha),$$

T. Dickhaus, *Simultaneous Statistical Inference*,
DOI: 10.1007/978-3-642-45182-9_2,

where the probability measure $\mathbb{P}^*$ is chosen such that

$$\mathbb{P}^*(T(X) \in \Gamma_\alpha) = \sup_{\vartheta \in H} \mathbb{P}_\vartheta(T(X) \in \Gamma_\alpha),$$

if H is a composite null hypothesis.

Remark 2.1.

(i) If H contains only one single element ϑ_0 (H is a simple hypothesis) and if $\mathbb{P}_H \equiv \mathbb{P}_{\vartheta_0}$ is continuous, it (typically) holds

$$p_\varphi(x) = \inf\{\alpha : T(x) \in \Gamma_\alpha\}.$$

(ii) In view of (3.3) in Lehmann and Romano (2005), we may regard the p-value as the "observed size" of φ.

(iii) Let Ω denote the domain of X. The mapping $p_\varphi(X) : \Omega \to [0, 1], \omega \mapsto p_\varphi(X(\omega))$, can be regarded as a random variable (under measurability assumptions). Often, there is no clear-cut distinction between the value $p_\varphi(x) \in [0, 1]$ and the random variable $p_\varphi(X)$. We will try to be as precise as possible with respect to this.

Definition 2.2. Under the assumptions of Definition 2.1, let the test statistic T fulfill the monotonicity condition

$$\forall \vartheta_0 \in H : \forall \vartheta_1 \in K : \forall c \in \mathbb{R} : \mathbb{P}_{\vartheta_0}(T(X) > c) \le \mathbb{P}_{\vartheta_1}(T(X) > c). \tag{2.1}$$

Then, we call φ a test of (generalized) Neyman-Pearson type, if for all $\alpha \in (0, 1)$ there exists a constant c_α, such that

$$\varphi(x) = \begin{cases} 1, & T(x) > c_\alpha, \\ 0, & T(x) \le c_\alpha. \end{cases}$$

In practice, the constants c_α are determined via $c_\alpha = \inf\{c \in \mathbb{R} : \mathbb{P}^*(T(X) > c) \le \alpha\}$ with $\mathbb{P}^*$ as in Definition 2.1 ("at the boundary of the null hypothesis"). If H is simple and $\mathbb{P}_H$ continuous, we obtain $c_\alpha = F_T^{-1}(1-\alpha)$, where F_T denotes the cdf. of $T(X)$ under H.

Lemma 2.1. *Let φ a test of Neyman-Pearson type and assume that $\mathbb{P}^*$ does not depend on α. Then it holds*

$$p_\varphi(x) = \mathbb{P}^*(T(X) \ge t^*) \text{ with } t^* = T(x).$$

Proof. The rejection regions $\Gamma_\alpha = (c_\alpha, \infty)$ are nested. Therefore, $\inf\{\alpha : T(x) \in \Gamma_\alpha\}$ is attained in $[t^*, \infty)$. The assertion follows from Definition 2.1. □

If H is simple, $\mathbb{P}_H$ continuous, and φ of Neyman-Pearson type, Lemma 2.1 yields $p_\varphi(x) = 1 - F_T(t^*)$, with F_T as in Definition 2.2.

Theorem 2.1 (p-values as decision tools). *Let $\alpha \in (0, 1)$ a fixed given significance level and assume that $\mathbb{P}^*$ is continuous. Then we have the duality*

$$\varphi(x) = 1 \Longleftrightarrow p_\varphi(x) < \alpha.$$

Proof. We restrict the proof to the case of tests of Neyman-Pearson type. The mapping $t \mapsto \mathbb{P}^*(T(X) > t)$ is decreasing in t. Moreover, due to the construction of c_α (see Definition 2.2), we must have $\mathbb{P}^*(T(X) > c_\alpha) \leq \alpha$ and $\mathbb{P}^*(T(X) > c) > \alpha$ for all $c < c_\alpha$. Altogether, this entails that $p_\varphi(x) < \alpha$ is equivalent to $t^* > c_\alpha$. The latter event characterizes rejection of H according to Definition 2.2. □

Remark 2.2.

(i) The advantage of p-values for testing is that they can be computed without prior specification of a significance level α. This is why all common statistics software systems implement statistical tests via the computation of p-values. However, for the purpose of decision making, pre-specification of α is inevitable.
(ii) The p-value gives an answer to the question "How probable are the observed data, given that the null hypothesis is true?". However, it does *not* answer the question "How probable is the validity of the null hypothesis, given the observed data?".
(iii) For some applications, it is more useful to consider isotone transformations of test statistics rather than antitone ones. Therefore, we remark here that $1 - p_\varphi(X)$ is in the cases that are relevant for our work equal to the *distributional transform* of $T(X)$ as defined by Rüschendorf (2009). We will adopt this terminology in the remainder of this work.

Theorem 2.2. *Under the assumptions of Definition 2.1, assume that H is simple, $\mathbb{P}_H$ is continuous and φ is a test of Neyman-Pearson type. Then it follows*

$$p_\varphi(X) \underset{H}{\sim} UNI[0, 1].$$

Proof. The assertion is a consequence of the principle of quantile transformation. Making use of Lemma 2.1, we easily calculate

$$\begin{aligned}
\mathbb{P}_H(p_\varphi(X) \leq t) &= \mathbb{P}_H(1 - F_T(T(X)) \leq t) \\
&= \mathbb{P}_H(F_T(T(X)) \geq 1 - t) \\
&= \mathbb{P}(U \geq 1 - t) = 1 - \mathbb{P}(U \leq 1 - t) \\
&= 1 - (1 - t) = t,
\end{aligned}$$

where U denotes a standard uniform variate. □

Remark 2.3. In general, it holds that $p_\varphi(X)$ is under H stochastically not smaller than a standard uniform variate, i.e.,

$$\forall \vartheta \in H : \mathbb{P}_\vartheta(p_\varphi(X) \leq t) \leq t, \; t \in [0, 1]. \tag{2.2}$$

Occasionally, p-values are even defined via property (2.2) in the literature, without reference to test statistics or rejection regions at all; see, for instance, Definition 8.3.26 in the textbook by Casella and Berger (2002).

2.1 Randomized p-values

In Theorem 2.2, we assumed a simple null hypothesis H and that $\mathbb{P}_H$ is continuous. Hence, two potential sources of non-uniformity of p-values are discreteness of $\mathbb{P}_H$ and testing of composite null hypotheses. In this section, we demonstrate how randomization techniques can be used to remove or at least to diminish the conservativity that we have reported in (2.2) if the statistical model entails one of the aforementioned sources of non-uniformity of the p-values in the sense of Definition 2.1. As we will point out later, this is important for multiple testing, especially because many data-adaptive multiple tests require exactly uniformly distributed p-values under null hypotheses for a reasonable performance and fail to work properly if this assumption is violated.

2.1.1 Randomized p-values in Discrete Models

We started with non-randomized tests in Definition 2.1. Especially in discrete models, this leads to p-values that are stochastically larger than UNI[0, 1]. To meet the requirement of uniformity of the p-values under null hypotheses at least for the case of testing point hypotheses, p-values can be slightly modified in analogy to randomization of tests.

Definition 2.3 (Realized randomized p-value). Let a statistical model $(\mathcal{X}, \mathcal{F}, (\mathbb{P}_\vartheta)_{\vartheta \in \Theta})$ be given. Consider the two-sided test problem $H : \{\vartheta = \vartheta_0\}$ versus $K : \{\vartheta \neq \vartheta_0\}$ and assume the decision is based on the realization x of a discrete random variate $X \sim \mathbb{P}_\vartheta$ with values in $\mathcal{X}$. Moreover, let U denote a uniformly distributed random variable on [0, 1], stochastically independent of X. Then, a realized randomized p-value for testing H versus K is a measurable mapping $p^{\text{rand.}} : \mathcal{X} \times [0, 1] \to [0, 1]$ fulfilling that

$$\mathbb{P}_{\vartheta_0}(p^{\text{rand.}}(X, U) \leq t) = t \text{ for all } t \in [0, 1]. \tag{2.3}$$

The property (2.3) is an abstract mathematical requirement. For practical applications, the following theorem which is due to Klaus Straßburger makes the concept of realized randomized p-values fully usable. The proof of Theorem 2.3 can be found in Appendix II of Dickhaus et al. (2012).

Theorem 2.3. *Let $T : \mathcal{X} \to \mathbb{R}$ denote a statistic and let $f : \mathcal{X} \to \mathbb{R}_+$ be the pmf. of a discrete random variate X with values in $\mathcal{X}$, such that $f(x) > 0$ for all $x \in \mathcal{X}$. Moreover, let U denote a UNI[0, 1]—distributed variate which is stochastically independent of X. Define*

$$p_T(x) = \sum_{y:T(y)\leq T(x)} f(y), \quad \mathcal{W} = \{p_T(x) : x \in \mathcal{X}\}, \text{ and}$$

$$p_T^{\text{rand.}}(x,u) = p_T(x) - u \sum_{y:T(y)=T(x)} f(y). \tag{2.4}$$

Then it holds

$$\begin{aligned} \mathbb{P}(p_T(X) \leq t) &\leq t, \text{ for all } t \in [0,1], \\ \mathbb{P}(p_T(X) \leq t) &= t, \text{ for all } t \in \mathcal{W}, \\ \mathbb{P}(p_T^{\text{rand.}}(X,U) \leq t) &= t, \text{ for all } t \in [0,1]. \end{aligned} \tag{2.5}$$

If realized randomized p-values are constructed according to (2.4), the relationship between p-value and distributional transform given in part (iii) of Remark 2.2 remains to hold.

2.1.2 Randomized p-values for Testing Composite Null Hypotheses

Dickhaus (2013) proposed randomized p-values for testing composite null hypotheses as follows.

Definition 2.4 (Dickhaus (2013)). Let p^{LFC} be a p-value which is constructed as in Definition 2.1 and let u denote the realization of a UNI[0, 1]—distributed random variable U which is stochastically independent of X. Then, the randomized p-value $p^{\text{rand.}}$ is given by

$$p^{\text{rand.}}(x,u) = u\mathbf{1}_H(\hat{\theta}(x)) + G(p^{\text{LFC}}(x))\mathbf{1}_K(\hat{\theta}(x)),$$

where $\theta : \Theta \to \Theta'$ denotes a one-dimensional (possibly derived) parameter, $\hat{\theta}$ a consistent and (at least asymptotically for large sample sizes) unbiased estimator of θ, and G the conditional cdf of $p^{\text{LFC}}(X)$ given $\hat{\theta} \in K$ under the (or: any) LFC for the type I error probability of the test φ of $H \subset \Theta'$ versus $K = H \setminus \Theta'$ corresponding to p^{LFC}.

At least for one-sided tests of means in Gaussian models, Dickhaus (2013) showed that these p-values are valid and under null hypotheses stochastically not larger than the traditional, LFC-based ones. Alternative methods for multiple testing of composite null hypotheses are reviewed in the introduction of Dickhaus (2013).

2.2 *p*-value Models

In the context of multiple test problems, (marginal) p-values $p_1, \ldots, p_m$ can be computed for every individual pair of hypotheses H_i versus K_i, if marginal models can, at least under null hypotheses, be specified exactly (which is often a hard requirement). A broad class of multiple tests depend on the data only via $p_1, \ldots, p_m$ and combine them in a suitable way in order to control errors, based on probabilistic calculations. Hence, for the mathematical analysis of such multiple tests, it suffices to model the distribution of the vector $(p_1(X), \ldots, p_m(X))^\top$ of (random) p-values and to consider statistical models of the form $([0, 1]^m, \mathscr{B}([0, 1]^m), (\mathbb{P}_\vartheta : \vartheta \in \Theta))$. Especially in high-dimensional settings, often only qualitative assumptions on the joint distribution of $p_1, \ldots, p_m$ (regarded as random variables) are made which lead to a variety of standard p-value models which are frequently considered in multiple hypotheses testing.

2.2.1 The iid.-Uniform Model

If one can assume that all m p-values $p_1, \ldots, p_m$ are stochastically independent and that the marginal test problems H_i versus K_i, $1 \leq i \leq m$, are such that Theorem 2.2 applies for all of them, then the joint distribution of $p_1(X), \ldots, p_m(X)$ under the global hypothesis H_0 is fully specified, because under these assumptions $p_1(X), \ldots, p_m(X)$ are under H_0 distributed as a vector $(U_1, \ldots, U_m)^\top$ of m stochastically independent, identically UNI[0, 1]—distributed random variables. Moreover, if only (without loss of generality) hypotheses $H_1, \ldots, H_{m_0}$ are true for some $m_0 = m_0(\vartheta) \in \{1, \ldots, m\}$, then $p_1(X), \ldots, p_{m_0}(X)$ are distributed as $(U_1, \ldots, U_{m_0})^\top$. We call this p-value model the iid.-uniform model.

For certain classes of multiple test procedures, the iid.-uniform model already implies the distribution of V_m and hence suffices to calibrate such multiple tests with respect to FWER control. To illustrate this, assume that the multiple test φ is such that those hypotheses are rejected for which the corresponding p-value is smaller than some given threshold $\alpha_{\text{loc.}} \in (0, 1)$. We call $\alpha_{\text{loc.}}$ a local significance level and such multiple test procedures single-step tests, cf. Sect. 3.1.1. Then, assuming the iid.-uniform model for $p_1(X), \ldots, p_m(X)$, V_m is under $\vartheta \in \Theta$ binomially distributed with parameters $m_0(\vartheta)$ and $\alpha_{\text{loc.}}$. This leads to the following expression for the FWER of φ under ϑ.

$$\mathrm{FWER}_\vartheta(\varphi) = \mathbb{P}_\vartheta(V_m > 0) = 1 - \mathbb{P}_\vartheta(V_m = 0) = 1 - (1 - \alpha_{\mathrm{loc.}})^{m_0}. \tag{2.6}$$

Obviously, the right-hand side of (2.6) is increasing in m_0. Therefore, under the iid.-uniform model, the FWER of a single-step test φ becomes largest for such ϑ for which $I_0(\vartheta) = I = \{1, \ldots, m\}$. In other words, all $\vartheta \in H_0$ are least favorable for the FWER of φ under the iid.-uniform model. This allows for a precise calibration of $\alpha_{\mathrm{loc.}}$ for strong FWER control (which is equivalent to weak FWER control here). The resulting single-step test is known as Šidák test and will be presented in Example 3.2. Since the full joint distribution of $(p_i(X) : i \in I_0(\vartheta))$ is completely specified in the iid.-uniform model, also the joint and the marginal distributions of the order statistics of the latter sub-vector of p-values can be derived and expressed in closed form. These distributions are important for calibrating step-up-down multiple test procedures. Such multiple tests reject hypotheses whose p-values are below a threshold which is determined data-dependently by the value of an order statistic of $(p_i(X) : 1 \leq i \leq m)$, see Sect. 3.1.2. Let us briefly recall the following facts.

Lemma 2.2. *Let $Y_1, \ldots, Y_m$ denote stochastically independent, identically distributed random variables driven by the probability measure $\mathbb{P}$, with cdf. F of Y_1. Assume that $\mathbb{P}^{Y_1}$ is absolutely continuous with respect to the Lebesgue measure λ and denote the order statistics of $(Y_1, \ldots, Y_m)^\top$ by $(Y_{1:m}, \ldots, Y_{m:m})^\top$. Then the following assertions hold true.*

$$\mathbb{P}(Y_{i:m} \leq y) = \sum_{j=i}^{m} \binom{m}{j} F(y)^j (1 - F(y))^{m-j},$$

$$\frac{d\mathbb{P}^{Y_{i:m}}}{d\mathbb{P}^{Y_1}}(y) = m\binom{m-1}{i-1} F(y)^{i-1}(1 - F(y))^{m-i}.$$

If $\mathbb{P}^{Y_1}$ has Lebesgue density f, then $\mathbb{P}^{Y_{i:m}}$ has Lebesgue density $f_{i:m}$, given by

$$f_{i:m}(y) = m\binom{m-1}{i-1} F(y)^{i-1}(1 - F(y))^{m-i} f(y). \tag{2.7}$$

Letting $\mu = \mathbb{P}^{Y_1}$, $(Y_{i:m})_{1 \leq i \leq m}$ has joint μ^m-density

$$(y_1, \ldots, y_m) \mapsto m!\mathbf{1}_{\{y_1 < y_2 < \cdots < y_m\}}.$$

If μ has Lebesgue density f, then $(Y_{i:m})_{1 \leq i \leq m}$ has λ^m-density

$$(y_1, \ldots, y_m) \mapsto m! \prod_{i=1}^{m} f(y_i)\mathbf{1}_{\{y_1 < y_2 < \cdots < y_m\}}.$$

Remark 2.4. Considering iid. UNI[0, 1]—distributed random variables $U_1, \ldots, U_m$ in Lemma 2.2, Eq. (2.7) shows that the order statistic $U_{i:m}$ has a Beta$(i, m - i + 1)$ distribution with

$$\mathbb{E}[U_{i:m}] = \frac{i}{m+1}, \quad \mathrm{Var}(U_{i:m}) = \frac{i(m-i+1)}{(m+1)^2(m+2)}.$$

For computing the joint cumulative distribution function of $(U_{1:m}, \ldots, U_{m:m})$, efficient recursive algorithms exist, for instance Bolshev's recursion and Steck's recursion (see Shorack and Wellner (1986), p. 362 ff.).

Lemma 2.2 can be used to calibrate a step-up-down multiple test procedure φ for weak FWER control under the assumption of an iid.-uniform model for the p-values $p_1(X), \ldots, p_m(X)$. However, if $\vartheta \notin H_0$, the FWER of φ typically depends on the distribution of $(p_j(X) : j \in I_1(\vartheta))$, too. The same holds true for the FDR of φ, because the distribution of R_m certainly relies on that of $(p_j(X) : j \in I_1(\vartheta))$. This shows that some assumptions on the p-value distribution under alternatives are also needed to study the behavior of multiple tests operating on p-values, even for the sole purpose of type I error rate control according to Definition 1.2. A generalization of Steck's recursion to two populations has been derived by Blanchard et al. (2014). This generalization can for instance be used for calibrating multiple tests for FDR control if a fixed alternative p-value distribution is assumed and all m p-values are stochastically independent.

2.2.2 Dirac-Uniform Configurations

It seems that the term "Dirac-uniform configuration" was used for the first time by Finner and Roters (2001). A Dirac-uniform configuration is characterized by three distributional assumptions regarding the joint distribution of $(p_1, \ldots, p_m)$.

Definition 2.5 (Dirac-uniform configuration). The value of the parameter ϑ is called a Dirac-uniform configuration if the following three distributional properties hold.

1. All m_0 marginal p-values corresponding to true null hypotheses are stochastically independent and identically distributed as UNI[0, 1].
2. The random vector $(p_i(X) : i \in I_0(\vartheta))$ is stochastically independent of the random vector $(p_j(X) : j \in I_1(\vartheta))$.
3. For all $j \in I_1$, $p_j(X)$ follows a Dirac distribution with point mass 1 in zero, meaning that p_j is almost surely equal to zero.

Of course, in practice it is unrealistic to assume that effect sizes are so large that p-values are almost surely equal to zero under alternatives. Therefore, Dirac-uniform configurations are not useful for modeling real-life data. This is why we do not term them "models". They are technical devices for deriving upper bounds for the FWER or the FDR of multiple testing procedures. For the mathematical analysis of multiple tests under independence assumptions, Dirac-uniform configurations are important tools, because they are, for fixed m_0, often LFCs for the FWER and/or the FDR of multiple tests if p-values are independent. In particular, if φ is a step-up-down test, its

FWER and FDR typically become maximum if parameter values under alternatives are extreme in the sense that p-values under alternatives are as small as possible. As a consequence, control of the respective error rate by φ under Dirac-uniform configurations (which are LFCs) entails that φ controls the error rate also under all other (often more realistic) values of the parameter ϑ of the model.

Furthermore, analytic calculations for the FWER and the FDR are very straightforwardly possible under Dirac-uniform configurations, because it holds (almost surely) that $R_m = V_m + m_1$ for a stepwise rejective multiple test φ, if ϑ is a Dirac-uniform configuration. This is because the m_1 null hypotheses with indices in I_1 are almost surely rejected by φ due to their p-values which are almost surely equal to zero. Consequently, the joint distribution of V_m and R_m is already determined by that of V_m which in turn can be expressed in terms of the joint distribution of order statistics of m_0 iid. UNI[0, 1]—distributed random variables, and the respective Bolshev's or Steck's recursions suffice for the type I error calibration of φ. The latter reasoning will play an important role in Chap. 5 where we will provide more details.

2.2.3 Two-Class Mixture Models

In contrast to the models discussed before, two-class mixture models are often used as models for real-life data. They still have a tractable structure.

Definition 2.6. The joint distribution of $(p_1, \ldots, p_m)$ is called a two-class mixture model, if the following two properties hold.

1. All m_0 p-values corresponding to true null hypotheses are marginally distributed with cdf. F_0, where F_0 is stochastically lower-bounded by UNI[0, 1].
2. All m_1 p-values corresponding to false null hypotheses are marginally distributed with cdf. F_1.

At a first glance, this model seems very restrictive, because all p-values under null hypotheses share the same marginal distribution and the same holds true for all p-values under alternatives. However, the following trick, mentioned for instance by Genovese and Wasserman (2004) and Farcomeni (2007), considerably extends the applicability of two-class mixture models: Even if it can not be assumed that all $p_i(X)$ with $i \in I_1$ share the same marginal distribution, we may at least assume that their marginal cdfs all belong to a class $\{F_\xi : \xi \in \Xi\}$. In addition, we may be able to put a prior distribution ν on Ξ. If these two requirements are fulfilled, let F_1 be defined by $F_1(t) = \int_\Xi F_\xi(t)\nu(d\xi)$. In an analogous manner, one can proceed for constructing the marginal distribution function F_0 under null hypotheses, if the multiple test problem does not already imply a fixed marginal distribution of p-values under null hypotheses, for instance UNI[0, 1]. However, let us mention here that putting a prior on the parameter space under null hypotheses is problematic from the classical (frequentist) viewpoint toward statistics.

Notice that Definition 2.6 only specifies the marginal distributions of p-values. As far as the dependency structure in two-class mixture models is concerned, one often assumes weak dependency in the sense of Definition 5.2, meaning that the ecdfs of $(p_i : i \in I_0)$ and $(p_j : j \in I_1)$ converge for $m \to \infty$ to F_0 and F_1, respectively, in the Glivenko-Cantelli sense. This gives enough structure to the statistical model for an asymptotic analysis of the behavior of multiple tests operating on such p-values.

2.2.4 Copula Models Under Fixed Margins

The following well-known theorem provides a convenient way to separate the models for the marginal distributions of $p_1(X), \ldots, p_m(X)$ (which are, at least under null hypotheses, often already implied by the test problems H_i versus K_i, $1 \leq i \leq m$, see Theorem 2.2) from a model regarding the dependency structure among the p-values.

Theorem 2.4 (Sklar (1959, 1996)). *Let $Y = (Y_1, \ldots, Y_m)^\top$ denote a random vector with values in $\mathbb{R}^m$ and with joint cdf F_Y and marginal cdfs $F_{Y_1}, \ldots, F_{Y_m}$. Then there exists a function $C : [0, 1]^m \to [0, 1]$, called the copula of Y, such that for all $y = (y_1, \ldots, y_m)^\top \in \bar{\mathbb{R}}^m$, it holds*

$$F_Y(y) = C(F_{Y_1}(y_1), \ldots, F_{Y_m}(y_m)).$$

If all m marginal cdfs are continuous, then the copula C is unique.

According to Theorem 2.4, the dependency structure among $p_1(X), \ldots, p_m(X)$ can be modeled by modeling their copula. Furthermore, if Theorem 2.2 applies for all marginal test problems H_i versus K_i, $1 \leq i \leq m$, the copula of the p-values coincides under the global hypothesis H_0 with the cdf of $p_1(X), \ldots, p_m(X)$. The latter fact is extremely useful for constructing simultaneous test procedures based on p-values, cf. Sect. 4.4. In particular, parametric copula models can be used as regularized models for the dependency structure of $p_1(X), \ldots, p_m(X)$, especially in cases where m is large such that the "curse of dimensionality" prohibits modeling or reliably estimating the full joint distribution of the data or the p-values, respectively. Regularization here means that the copula parameter is of low dimension. Of course, in practice this will typically only yield an approximation of the true dependency structure.

2.2.5 Further Joint Models

Assume that all marginal tests φ_i, $1 \leq i \leq m$, for a given multiple test problem $(\mathcal{X}, \mathcal{F}, (\mathbb{P}_\vartheta)_{\vartheta \in \Theta}, \mathcal{H}_m)$ are of (generalized) Neyman-Pearson type in the sense of Definition 2.2, with (marginal) test statistics $T_1, \ldots, T_m$. Then, in order to calibrate the multiple test φ by multivariate techniques, it is often convenient to consider the joint distribution of $T_1, \ldots, T_m$ directly. On the p-value scale, however, the resulting adjustment for multiplicity of the overall significance level α (for the FWER or the FDR) is explicitly given, leading to a better interpretability of φ. Therefore, it may be of interest to derive the joint distribution of the p-values $p_1, \ldots, p_m$ corresponding to $T_1, \ldots, T_m$ by transformation of measures. To give a specific example, assume that $X = (X_1, \ldots, X_m)^\top$ follows a multivariate normal distribution with mean $\mu = (\mu_1, \ldots, \mu_m)^\top$ and covariance matrix Σ. For ease of exposition and without loss of generality, assume that all diagonal elements of Σ are equal to one. Furthermore, assume that the m null hypotheses $H_i : \{\mu_i = 0\}$ with two-sided alternatives $K_i : \{\mu_i \neq 0\}$ are of interest, $1 \leq i \leq m$. Suitable test statistics are given by $T_i = |X_i|$, $1 \leq i \leq m$. Following Lemma 2.1 and utilizing symmetry properties of the standard normal law, the marginal p-values corresponding to the test statistics $T_1, \ldots, T_m$ are given by

$$p_i(x) = 2(1 - \Phi(T_i(x))), \quad 1 \leq i \leq m, \tag{2.8}$$

where Φ denotes the cdf. of the standard normal distribution.

Hence, if the calibration of φ results in a threshold c_α for the T_i, then equivalently φ_i rejects H_i if $p_i(x) < 2(1 - \Phi(c_\alpha)) = \alpha_{\text{loc.}}$ (say). The value $\alpha_{\text{loc.}}$ can thus be regarded as a multiplicity-adjusted local significance level.

Under H_i, $p_i(X)$ is marginally UNI[0, 1]—distributed, see Theorem 2.2. Moreover, the joint cdf of $(p_i(X) : 1 \leq i \leq m)$ under μ and Σ is given by

$$\begin{aligned} u = (u_1, \ldots, u_m)^\top \in [0, 1]^m \mapsto & \ \mathbb{P}_{(\mu,\Sigma)}(p_1(X) \leq u_1, \ldots, p_m(X) \leq u_m) \\ & = \mathbb{P}_{(\mu,\Sigma)}(\forall 1 \leq i \leq m : T_i \geq \Phi^{-1}(1 - u_i/2)). \end{aligned}$$

The latter probability can easily be computed by employing numerical routines for multivariate normal distributions, cf. Genz and Bretz (2009). Multiple tests for Gaussian means play an important role in many practical applications, for instance in the context of localized comparisons in analysis of variance models. Applications in genetics are discussed in Chaps. 9 and 10.

Acknowledgments Parts of Sect. 2.1 originated from joint work with Klaus Straßburger. I am grateful to Mette Langaas and Øyvind Bakke for inviting me and for their hospitality during my visit to Norwegian University of Science and Technology (NTNU), for many fruitful discussions and for critical reading.

References

Blanchard G, Dickhaus T, Roquain E, Villers F (2014) On least favorable configurations for step-up-down tests. Statistica Sinica 24(1):1–23

Casella G, Berger RL (2002) Statistical inference, 2nd edn. Brooks/Cole, Cengage Learning

Dickhaus T (2013) Randomized p-values for multiple testing of composite null hypotheses. J Stat Plann Infer 143(11):1968–1979

Dickhaus T, Strassburger K, Schunk D, Morcillo-Suarez C, Illig T, Navarro A (2012) How to analyze many contingency tables simultaneously in genetic association studies. Stat Appl Genet Mol Biol 11(4):Article 12

Farcomeni A (2007) Some results on the control of the false discovery rate under dependence. Scand J Stat 34(2):275–297. doi:10.1111/j.1467-9469.2006.00530.x

Finner H, Roters M (2001) On the false discovery rate and expected type I errors. Biom J 43(8):985–1005

Genovese C, Wasserman L (2004) A stochastic process approach to false discovery control. Ann Stat 32(3):1035–1061. doi:10.1214/009053604000000283

Genz A, Bretz F (2009) Computation of multivariate normal and t probabilities. Lect Notes Stat 195. Springer, Berlin. doi:10.1007/978-3-642-01689-9

Lehmann EL, Romano JP (2005) Testing statistical hypotheses, 3rd ed. Springer Texts in Statistics, New York

Rüschendorf L (2009) On the distributional transform, Sklar's theorem, and the empirical copula process. J Stat Plann Inference 139(11):3921–3927. doi:10.1016/j.jspi.2009.05.030

Shorack GR, Wellner JA (1986) Empirical processes with applications to statistics. Wiley Series in Probability and Mathematical Statistics. Wiley, New York

Sklar A (1959) Fonctions de répartition à n dimensions et leurs marges. Publ Inst Stat Univ Paris 8:229–231

Sklar A (1996) Random variables, distribution functions, and copulas—a personal look backward and forward. In: Distributions with fixed marginals and related topics, IMS Lecture Notes-Monograph Series, Volume 28, Institute of Mathematical Statistics, Hayward, CA, pp 1–4

Chapter 3
Classes of Multiple Test Procedures

Abstract The aim of this chapter is a systematic overview of different classes of multiple tests. Procedures are distinguished by their structure, by the degree of detail of the underlying statistical model and by the type of error control that they provide. Major categories comprise margin-based multiple tests, multivariate multiple test procedures and closed test procedures. Subcategories are introduced where appropriate. We discuss specific examples and indicate computer implementations by means of flow diagrams and pseudo-code. Applications and references to later chapters illustrate which kind of multiple test procedure can be utilized for some standard types of multiple test problems which are relevant in practice. Precise references to the literature are collected for a deeper study of specific methods.

Although the literature on multiple test procedures (MTPs) is nowadays exponentially increasing over time, it is still possible to systematize the proposed methods according to some general categories. For instance, one class of methods only models the marginal distributions of the involved test statistics explicitly and combines these test statistics or, equivalently, corresponding p-values following probabilistic calculations. We call resulting procedures margin-based multiple test procedures. Different margin-based MTPs employ different qualitative assumptions on the dependency structure between test statistics or p-values, cf. our Chap. 2. Examples of this kind of procedures are discussed in Sect. 3.1.

Another class of MTPs considers the full joint distribution of all test statistics and relies on calculating or approximating quantiles of this joint distribution, for instance by resampling or by proving asymptotic normality by means of central limit theorems. We term such procedures multivariate multiple test procedures and discuss them in Sect. 3.2. A class of in a certain sense hybrid (neither purely margin-based nor entirely multivariate) multiple test procedures, which are specifically tailored to control the FWER in structured systems of hypotheses, is constituted by closed test procedures, which we will treat in Sect. 3.3.

Further criteria to distinguish MTPs are their structure (single-step or stepwise rejective), and the type of error control (k-FWER-controlling, FDR-controlling, FDX-controlling, etc.) that they provide. We exclude a distinction between frequentist

T. Dickhaus, *Simultaneous Statistical Inference*,
DOI: 10.1007/978-3-642-45182-9_3,

and Bayesian procedures here, because this work is not considered with Bayesian approaches to multiple hypotheses testing. As far as frequentist procedures are concerned, the aforementioned criteria in our opinion allow us to treat the majority of the most popular MTPs up to present.

One type of procedures which do not fit in a clear-cut way into the categories defined above is constituted by so-called augmentation procedures. Augmentation procedures for control of the k-FWER, the FDR or the FDX work in two stages: In the first stage, an FWER-controlling MTP is applied. In the second stage, a certain number of hypotheses not rejected by the procedure employed in the first stage is rejected additionally, whereby this number in general depends on the data and on probabilistic bounds. Although augmentation procedures have attracted some attention recently, we do not cover them in the present work. References for augmentation procedures include van der Laan et al. (2004; 2005), and Farcomeni (2009).

3.1 Margin-Based Multiple Test Procedures

The multiple tests discussed in this section only require that each marginal test φ_i can be calibrated to keep a local significance level $\alpha_{\text{loc.}}$ (say). The multiple test $\varphi = (\varphi_i : 1 \leq i \leq m)$ is then built up from these marginal tests by adjusting $\alpha_{\text{loc.}}$ for the multiplicity of the problem. This adjustment may be given by an explicit "correction for multiplicity" based on probabilistic considerations or in a data-dependent manner, for instance by defining $\alpha_{\text{loc.}}$ by the value of an order statistic of marginal p-values $p_1, \ldots, p_m$.

3.1.1 Single-Step Procedures

Single-step multiple test procedures carry out each individual test φ_i, $1 \leq i \leq m$, at (local) significance level $\alpha_{\text{loc.}}$, where $\alpha_{\text{loc.}}$ is the result of a multiplicity correction of α. In view of Theorem 2.1, single-step multiple tests are extremely easy to carry out in practice: Just calculate marginal p-values $p_1, \ldots, p_m$ and reject H_i if and only if $p_i < \alpha_{\text{loc.}}$. The choice of $\alpha_{\text{loc.}}$ depends on qualitative assumptions regarding the joint distribution of $(p_1, \ldots, p_m)$. Two classical procedures are the Bonferroni correction (or Bonferroni test) and the Šidák correction (or Šidák test).

Example 3.1 (Bonferroni correction, cf. Bonferroni (1935; 1936)). The Bonferroni correction is based on the union bound and consists in choosing $\alpha_{\text{loc.}} = \alpha/m$. It provides strong control of the FWER without any assumptions on the dependency structure among $(p_1, \ldots, p_m)$, because for a Bonferroni test φ, it holds for all $\vartheta \in \Theta$ that

$$\begin{aligned}\text{FWER}_\vartheta(\varphi) &= \mathbb{P}_\vartheta\Big(\bigcup_{i\in I_0(\vartheta)}\{\varphi_i = 1\}\Big)\\ &\le \sum_{i\in I_0(\vartheta)} \mathbb{P}_\vartheta(\{\varphi_i = 1\})\\ &\le m_0\alpha/m \le \alpha.\end{aligned}$$

The inequality $\mathbb{P}(\bigcup_{i=1}^m A_i) \le \sum_{i=1}^m \mathbb{P}(A_i)$ is referred to as Bonferroni inequality in the multiple testing literature.

The disadvantage of Bonferroni tests is that α/m is very small for large m. Therefore, Bonferroni tests have low multiple power if m is large. If joint independence of all m marginal p-values can be assumed, $\alpha_{\text{loc.}}$ can be chosen slightly larger than α/m.

Example 3.2 (Šidák correction, cf. Šidák 1967). The Šidák correction consists in choosing $\alpha_{\text{loc.}} = 1 - (1-\alpha)^{1/m}$. It provides strong control of the FWER if $(p_1, \ldots, p_m)$ are jointly stochastically independent, because for a Šidák test φ, it then holds for all $\vartheta \in \Theta$ that

$$\begin{aligned}\text{FWER}_\vartheta(\varphi) &= \mathbb{P}_\vartheta\Big(\bigcup_{i\in I_0(\vartheta)}\{\varphi_i = 1\}\Big)\\ &= 1 - \mathbb{P}_\vartheta\Big(\bigcap_{i\in I_0(\vartheta)}\{\varphi_i = 0\}\Big)\\ &= 1 - \prod_{i\in I_0(\vartheta)} \mathbb{P}_\vartheta(\{\varphi_i = 0\})\\ &\le 1 - \prod_{i\in I_0(\vartheta)} (1-\alpha)^{1/m}\\ &= 1 - (1-\alpha)^{m_0/m}\\ &\le 1 - (1-\alpha) = \alpha.\end{aligned}$$

As mentioned before, for all $m \in \mathbb{N}$ it holds $\alpha/m < 1-(1-\alpha)^{1/m}$, so that the more restrictive model assumptions made for a Šidák test allow one to increase multiple power uniformly. We may remark here that Šidák tests control the FWER under certain forms of positive dependence among $(p_1, \ldots, p_m)$, too. More details are provided in Chap. 4. Also asymptotically, it holds $m[1-(1-\alpha)^{1/m}] \to -\ln(1-\alpha) > \alpha = m\alpha/m$, $m \to \infty$, for any $\alpha \in (0, 1)$. However, also for the Šidák correction, we have $\alpha_{\text{loc.}} \to 0$, $m \to \infty$.

In the particular context of testing linear contrasts in Gaussian models, Scheffé (1953) obtained the following result.

Theorem 3.1 (Scheffé (1953)). *Let $k \ge 3$ and $n_i \ge 2$ for all $1 \le i \le k$ be given integers and $X = (X_{ij} : 1 \le i \le k, 1 \le j \le n_i)$. Assume that all X_{ij} are stochastically independent and normally distributed, $X_{ij} \sim \mathcal{N}(\mu_i, \sigma^2)$, where*

$\mu_i \in \mathbb{R}$, $1 \leq i \leq k$, and $\sigma^2 > 0$. For notational convenience, denote $n. = \sum_{i=1}^{k} n_i$. Consider the linear subspace

$$\mathscr{L} = \left\{ \sum_{j=1}^{q} h_j a^{(j)} \right\}$$

of $\mathbb{R}^k$ of dimension $q \leq k$, where $h_j \in \mathbb{R}$ for all $1 \leq j \leq q$ and $a^{(1)}, \ldots, a^{(q)} \in \mathbb{R}^k$ are linearly independent vectors. Then it holds for all $\mu \in \mathbb{R}^k$ and for all $\sigma^2 > 0$ that

$$\mathbb{P}_{(\mu,\sigma^2)} \left(\forall c \in \mathscr{L} : c^T \mu \in \left[c^T \hat{\mu} \mp \sqrt{q \widehat{Var}(c^T \hat{\mu}) F_{q,n.-k;\alpha}} \right] \right) = 1 - \alpha, \qquad (3.1)$$

where $\mu = (\mu_1, \ldots, \mu_k)^\top$, $\hat{\mu} = (\overline{X}_{1.}, \ldots, \overline{X}_{k.})^T$ (vector of empirical group means), and $\widehat{Var}(c^T \hat{\mu}) = s^2 \sum_{i=1}^{k} (c_i^2 / n_i)$, with s^2 denoting the pooled unbiased estimator of σ^2, and $F_{q,n.-k;\alpha}$ the upper α-quantile of Fisher's F-distribution with q and $n. - k$ degrees of freedom.

Equation (3.1) yields a simultaneous $1-\alpha$ confidence region for *all* linear contrasts of group means defined by $\mathscr{L}$ in the considered analysis of variance model. By duality of tests and confidence regions (see Theorem 1.1), this also entails a multiple single-step test for such contrasts.

3.1.2 Stepwise Rejective Multiple Tests

An interesting other class of multiple test procedures are stepwise rejective tests. In contrast to single-step tests, here the hypotheses are ordered by a pre-defined criterion and tested one after the other, where testing can stop at every step due to the occurrence of a rejection or a non-rejection. This means that the test result for a particular pair of hypotheses H_i versus K_i depends on the data not only directly via the test statistic T_i or the p-value p_i, but also indirectly via potentially all other test statistics or p-values. The way the ordering among the hypotheses is defined leads to different subtypes of stepwise rejective multiple tests.

3.1.2.1 Step-Up-Down Tests

Step-up-down tests, introduced by Tamhane et al. (1998), rely on an ordering of the hypotheses $H_1, \ldots, H_m$ which is induced by the order statistics of marginal p-values $p_1, \ldots, p_m$.

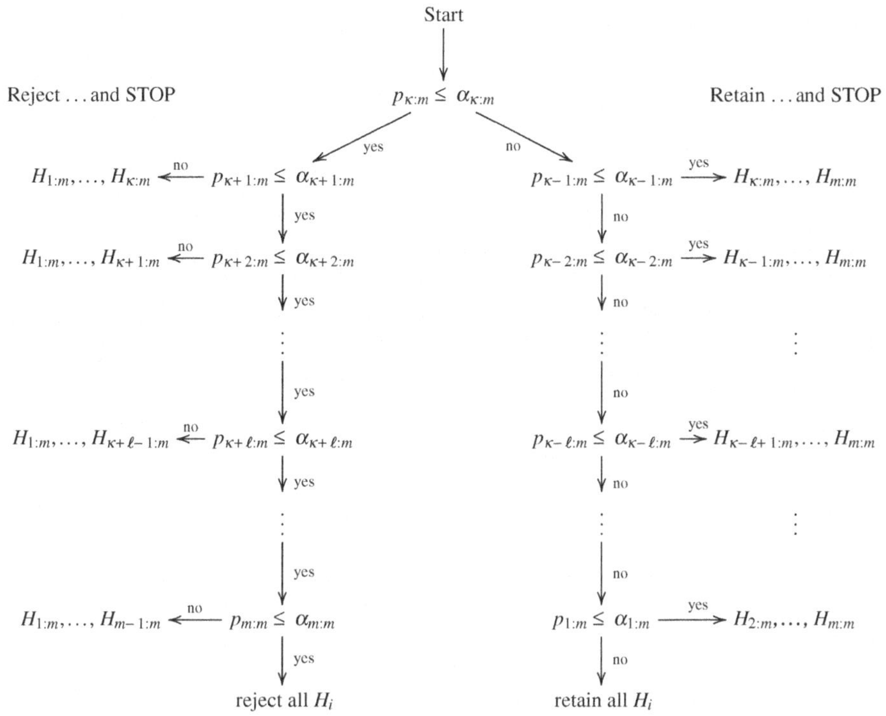

Fig. 3.1 Decision rule of an SUD test. If $\kappa = m$ (SU test) and $p_{m:m} \le \alpha_{m:m}$, all m null hypotheses are rejected. If $\kappa = 1$ (SD test) and $p_{1:m} > \alpha_{1:m}$, all m null hypotheses are retained

Definition 3.1 (Step-up-down test of order κ, cf. Finner et al. (2012)). Let $p_{1:m} < p_{2:m} < \cdots < p_{m:m}$ denote the ordered marginal p-values for a multiple test problem. For a tuning parameter $\kappa \in \{1, \ldots, m\}$ a step-up-down (SUD) test $\varphi^{\kappa} = (\varphi_1^{\kappa}, \ldots, \varphi_m^{\kappa})$ of order κ based on some critical values $\alpha_{1:m} \le \cdots \le \alpha_{m:m}$ is defined as follows. If $p_{\kappa:m} \le \alpha_{\kappa:m}$, set $j^* = \max\{j \in \{\kappa, \ldots, m\} : p_{i:m} \le \alpha_{i:m}$ for all $i \in \{\kappa, \ldots, j\}\}$, whereas for $p_{\kappa:m} > \alpha_{\kappa:m}$, put $j^* = \sup\{j \in \{1, \ldots, \kappa - 1\} : p_{j:m} \le \alpha_{j:m}\}$ ($\sup \emptyset = -\infty$). Define $\varphi_i^{\kappa} = 1$ if $p_i \le \alpha_{j^*:m}$ and $\varphi_i = 0$ otherwise ($\alpha_{-\infty:m} = -\infty$).

A step-up-down test of order $\kappa = 1$ or $\kappa = m$, respectively, is called step-down (SD) or step-up (SU) test, respectively. If all critical values are identical, we obtain a single-step test.

Figure 3.1 illustrates the decision rule of an SUD test schematically.

As we will discuss in Chap. 5, many commonly used step-up-down tests are margin-based and only employ qualitative assumptions regarding the joint distribution of test statistics or p-values. For instance, this holds true for the multiple tests by Holm (1979) (which are FWER-controlling step-down tests) and the famous linear step-up test by Benjamini and Hochberg (1995) for FDR control. However, there are remarkable exceptions, especially shortcuts of closed test procedures, cf. Sect. 3.3. The following obvious lemma can be used to compare different SUD tests which keep the same type I error criterion.

Lemma 3.1. *Consider two SUD tests* $\varphi^{(1)}$ *and* $\varphi^{(2)}$ *for the same multiple test problem* $(\mathscr{X}, \mathscr{F}, (\mathbb{P}_\vartheta)_{\vartheta \in \Theta}, \mathscr{H}_m)$. *Assume that one of the following properties holds true.*

(a) The two tests $\varphi^{(1)}$ *and* $\varphi^{(2)}$ *employ the same set of critical values and the tuning parameter* κ_2 *of* $\varphi^{(2)}$ *is larger than the tuning parameter* κ_1 *of* $\varphi^{(1)}$.
(b) The two tests $\varphi^{(1)}$ *and* $\varphi^{(2)}$ *employ the same tuning parameter* κ *and the critical values utilized in* $\varphi^{(2)}$ *are index-wise not smaller than the ones utilized in* $\varphi^{(1)}$.
(c) Both tests $\varphi^{(1)}$ *and* $\varphi^{(2)}$ *are single-step tests and the critical value utilized in* $\varphi^{(2)}$ *is larger than that utilized in* $\varphi^{(1)}$.

Then, for any realization of $(p_1, \ldots, p_m)^\top$, $\varphi^{(2)}$ *rejects all hypotheses that are rejected by* $\varphi^{(1)}$, *and possibly more.*

Hence, under the constraint of type I error control of given type and at given level, an optimal SUD test (with respect to multiple power, cf. Definition 1.4) is given by choosing κ and $\alpha_{1:m}, \ldots, \alpha_{m:m}$ as large as possible. For instance, SU tests have higher (not smaller) multiple power than the corresponding SD tests (with the same set of critical values). On the other hand, the same holds true for the comparison with respect to the FWER. Let us mention that additional assumptions are required in order that more rejections entail larger FDR, cf. Theorem 5.7.

Notice that we implicitly used part (c) of Lemma 3.1 for the comparison of Bonferroni tests and Šidák tests. In Chap. 5, Lemma 3.1 will be used for discussing relationships between the dependency structure among $p_1, \ldots, p_m$ and the choice of tuning parameters and critical values for SUD tests.

3.1.2.2 Fixed Sequence Multiple Tests

Similarly to step-up-down tests, fixed sequence multiple tests also rely on an ordering of the hypotheses $H_1, \ldots, H_m$. However, the ordering is now not data-dependently given by the ordering of p-values or test statistics, but is pre-defined before testing starts, for instance by weighting the hypotheses for importance. With respect to control of the FWER, the following fixed sequence procedure is widely used.

Theorem 3.2. Let $(\mathscr{X}, \mathscr{F}, \mathscr{P}, \mathscr{H})$ with $\mathscr{H} = (H_i : 1 \leq i \leq m)$ denote a multiple test problem and assume that valid marginal p-values $p_1, \ldots, p_m$ are at hand. Let $\alpha \in (0, 1)$ be a given constant and consider the multiple test φ defined by the following rule: Reject exactly hypotheses $H_1, \ldots, H_{k^*}$, where

$$k^* = \max\{1 \leq i \leq m : p_j \leq \alpha \text{ for all } j = 1, \ldots, i\}.$$

If k^* does not exist, retain all m null hypotheses. Then, φ strongly controls the FWER at level α.

Proof. First, consider the case $m = 2$. We have to distinguish four cases.

1. If both H_1 and H_2 are false, no type I error can occur, hence $\text{FWER}_\vartheta(\varphi) = 0$ for such ϑ.
2. If only H_1 is true, $\text{FWER}_\vartheta(\varphi) = \mathbb{P}_\vartheta(p_1(X) \leq \alpha) \leq \alpha$.
3. If only H_2 is true, $\text{FWER}_\vartheta(\varphi) = \mathbb{P}_\vartheta(\{p_1(X) \leq \alpha\} \cap \{p_2(X) \leq \alpha\}) \leq \mathbb{P}_\vartheta(p_2(X) \leq \alpha) \leq \alpha$.
4. If both H_1 and H_2 are true, $\text{FWER}_\vartheta(\varphi) = \mathbb{P}_\vartheta(p_1(X) \leq \alpha) \leq \alpha$.

It is easy to check that the latter reasoning remains to hold true for $m > 2$. □

The obvious drawback of the multiple test φ from Theorem 3.2 is that, once a particular hypothesis cannot be rejected, the remaining not yet rejected hypotheses have to be retained without being tested explicitly. Wiens (2003) developed a method based on a Bonferroni-type adjustment of α that allows for continuing testing after potential non-rejections. Other related testing strategies for fixed sequences of (pre-ordered) hypotheses ensuring strict FWER control have been discussed by Westfall and Krishen (2001) and Bauer et al. (1998), among many others. Such methods are particularly important for clinical trials with multiple endpoints.

3.1.3 Data-Adaptive Procedures

From the calculations in Examples 3.1 and 3.2, it follows that the realized k-FWER of the investigated margin-based multiple tests crucially depends on the proportion $\pi_0 = m_0/m$ of true null hypotheses. In Chap. 5, we will show that the same holds true for the realized FDR of many classical step-up-down tests. Data-adaptive procedures aim at adapting to the unknown quantity π_0 in order to exhaust the type I error level better and, consequently, increase multiple power of standard procedures. Explicitly adaptive (plug-in) procedures employ an estimate $\hat{\pi}_0$ and plug $\hat{\pi}_0$ into critical values, typically replacing m by $m \cdot \hat{\pi}_0$. In view of Definition 1.4 and Lemma 3.1, this increases multiple power at least on parameter subspaces on which $\mathbb{P}_\vartheta(\hat{\pi}_0 < 1)$ is large.

Maybe, the still most popular though, as well, the most ancient estimation technique for π_0 is the one of Schweder and Spjøtvoll (1982). It relies on a tuning parameter $\lambda \in [0, 1)$. Denoting the empirical cumulative distribution function (ecdf) of m marginal p-values by $\hat{F}_m$, the proposed estimator from Schweder and Spjøtvoll (1982) can be written as

$$\hat{\pi}_0 \equiv \hat{\pi}_0(\lambda) = \frac{1 - \hat{F}_m(\lambda)}{1 - \lambda}. \tag{3.2}$$

Among others, Storey et al. (2004), Langaas et al. (2005), Finner and Gontscharuk (2009), Dickhaus et al. (2012) and Dickhaus (2013) have investigated theoretical properties of $\hat{\pi}_0$ and slightly modified versions of this estimator. There exist several possible heuristic motivations for the usage of $\hat{\pi}_0$. The simplest one considers a

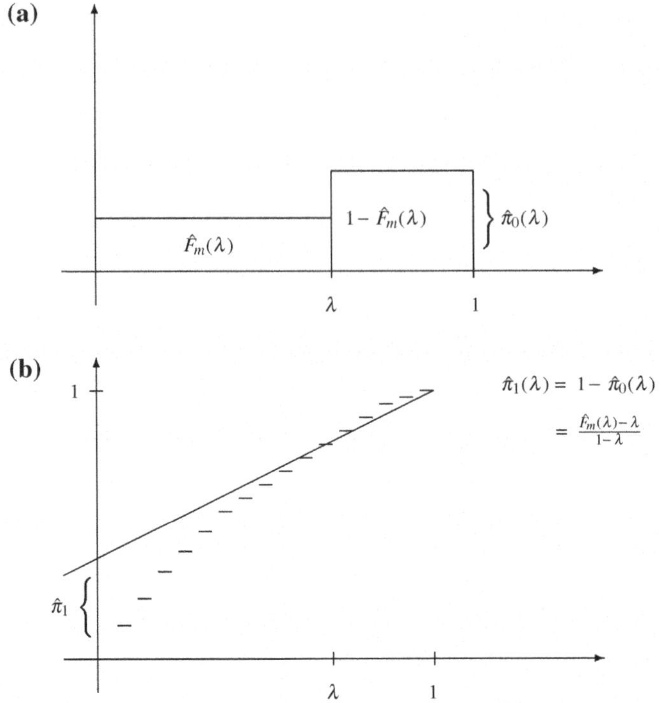

Fig. 3.2 Two graphical representations of the Schweder-Spjøtvoll estimator $\hat{\pi}_0(\lambda)$

histogram of the marginal p-values with exactly two bins, namely $[0, \lambda]$ and $(\lambda, 1]$. Then, the height of the bin associated with $(\lambda, 1]$ equals $\hat{\pi}_0(\lambda)$, see graph (a) in Fig. 3.2. A graphical algorithm for computing $\hat{\pi}_0$ connects the point $(\lambda, \hat{F}_m(\lambda))$ with the point $(1, 1)$. The offset of the resulting straight line at $t = 0$ equals $\hat{\pi}_1 = \hat{\pi}_1(\lambda) = 1 - \hat{\pi}_0(\lambda)$, see graph (b) in Fig. 3.2.

The following lemma is due to Dickhaus et al. (2012), see Lemma 1 in their paper.

Lemma 3.2. *Whenever* $(p_1, \ldots, p_m)$ *are valid p-values, i.e., marginally stochastically not smaller than* UNI$[0, 1]$ *under null hypotheses, the value of* $\hat{\pi}_0$ *is a conservative estimate of* π_0*, meaning that* $\hat{\pi}_0$ *has a non-negative bias. More specifically, it holds*

$$\mathbb{E}_\vartheta[\hat{\pi}_0(\lambda)] - \pi_0 \geq \frac{1}{m(1-\lambda)} \sum_{i \in I_1} \mathbb{P}_\vartheta(p_i > \lambda) \geq 0.$$

The data-adaptive Bonferroni plug-in (BPI) test by Finner and Gontscharuk (2009) replaces m by $m \cdot \hat{\pi}_0$ in the Bonferroni-corrected threshold for marginal p-values and the asymptotic version of the data-adaptive multiple test procedure by Storey et al. (2004) (STS test) replaces m by $m \cdot \hat{\pi}_0$ in Simes' critical values, cf. Sect. 5.3.

Another class of data-adaptive multiple tests is constituted by two-stage or multistage adaptive procedures, see Benjamini and Hochberg (2000) or Benjamini et al. (2006), for example. Such methods employ the number of rejections of a multiple test applied in the first stage in an estimator for m_0. This estimator is then used to calibrate the second stage test which leads to the actual decisions, where this principle may be applied iteratively. A third of class of methods is given by implicitly adaptive procedures. Here, the idea is to find critical values that automatically (for as many values of π_0 as possible) lead to full exhaustion of the type I error level. To this end, worst-case situations (i.e., LFCs) build the basis for the respective calculations. We will present some of such implicitly adaptive multiple tests in Sect. 5.5. Further estimation techniques for π_0 have also been proposed in the multiple testing literature. We defer the reader to the introduction in Finner and Gontscharuk (2009) for an overview.

3.2 Multivariate Multiple Test Procedures

The basic idea behind multivariate multiple test procedures is to incorporate the dependency structure of the data explicitly into the multiple test and thereby optimizing its power. The general reason why this is often possible is that margin-based procedures which control a specific multiple type I error rate have to provide this multiple type I error control generically over a potentially very large family of dependency structures. Hence, if it is possible to derive or to approximate the particular dependency structure for the data-generating distribution at hand, this information may be helpful to fine-tune a multiple test for this specific case. This is particularly important for applications from modern life sciences, because the data there are often spatially, temporally, or spatio-temporally correlated as we will demonstrate in later chapters. Three alternative ways to approximate dependency structures are resampling (Sect. 3.2.1), proving asymptotic normality by means of central limit theorems (Sect. 3.2.2), and fitting copula models (Sect. 3.2.3).

3.2.1 *Resampling-Based Methods*

It is fair to say that the basic reference for resampling-based FWER control is the book by Westfall and Young (1993), who introduced simultaneous and step-down multiple tests based on resampling under the assumption of subset pivotality (see Definition 4.3, basically meaning that the joint distribution of test statistics corresponding to true null hypotheses does not depend on the distribution of the remaining test statistics such that resampling under the global hypothesis H_0 is not only providing weak, but also strong FWER control). This assumption has been criticized as too restrictive such that (among others) Troendle (1995) and Romano and Wolf (2005a, b) generalized the methods of Westfall and Young (1993) to dispense with subset pivotality.

FDR-controlling (asymptotic) multiple tests based on resampling have been derived by Yekutieli and Benjamini (1999), Troendle (2000), and Romano et al. (2008). The resampling methods developed by Dudoit and van der Laan (2008) (see also the references therein) provide a general framework for controlling a variety of error rates (some of which we have introduced in Definitions 1.2 and 1.3), with particular emphasis on applications in genetics. While resampling often only asymptotically (for the sample size n tending to infinity) reproduces the true data distribution, Arlot et al. (2010) provide an in-depth study of resampling methods that control the FWER strictly for finite n.

3.2.2 Methods Based on Central Limit Theorems

Asymptotic normality of moment and maximum likelihood estimators are classical results in mathematical statistics, see, for instance, Chap. 12 by Lehmann and Romano (2005) or Chap. 5 by Van der Vaart (1998). We will discuss the special cases of multiple linear regression models and of generalized linear models in Chap. 4. If the vector T of test statistics for a given multiple test problem is (a transformation of) such an asymptotically normal point estimator, the asymptotic distribution of T can be derived and utilized for calibrating the multiple test. This has been demonstrated, for instance, by Hothorn et al. (2008) and Bretz et al. (2010) in general parametric models. For particular applications in genetic association studies (cf. Chap. 9), central limit theorems for multinomial distributions, together with positive dependency properties of multivariate chi-square distributions, have been exploited by Moskvina and Schmidt (2008) and Dickhaus and Stange (2013) (see also the references therein).

3.2.3 Copula-Based Methods

As discussed in Chap. 2, p-values are under certain assumptions uniformly distributed on $[0, 1]$ under null hypotheses. In particular, this holds true in many models which are typically used in life science applications. One example is the problem of multiple testing for differential gene expression, see Chap. 10. Hence, according to Theorem 2.4, in such cases it suffices to estimate the (often unknown) copula of $p_1(X), \ldots, p_m(X)$ in order to calibrate a multivariate multiple test procedure operating on these p-values. In particular, parametric copula models are convenient, because the dependency structure can in such models be condensed into a low-dimensional copula parameter. A flexible class of copula models is constituted by the family of Archimedean copulae.

Definition 3.2 (Archimedean copula). The joint distribution of the random vector $(p_i(X) : 1 \leq i \leq m)$ under $\vartheta \in \Theta$ is given by an Archimedean copula with copula generator ψ, if for all $(t_1, \ldots, t_m)^\top \in [0, 1]^m$,

$$\mathbb{P}_{\vartheta,\psi}(p_1(X) \leq t_1, \ldots, p_m(X) \leq t_m) = \psi\left(\sum_{i=1}^{m} \psi^{-1}\left(F_{p_i(X)}(t_i)\right)\right), \tag{3.3}$$

where $F_{p_i(X)}$ denotes the marginal cdf of $p_i(X)$ under $\vartheta \in \Theta$.

Dickhaus and Gierl (2013) demonstrated the usage of Archimedean copula models for FWER control, while Bodnar and Dickhaus (2013) are considered with FDR control under Archimedean p-value copulae. If the generator ψ only depends on a copula parameter η (say), standard parametric estimation approaches can be employed to estimate η. Two plausible estimation strategies are the maximum likelihood method (see, e. g., Hofert et al. (2012)) or the method of moments (referred to as "realized copula" method by Fengler and Okhrin (2012)). For the latter approach, the "inversion formulas" provided in the following lemma are helpful.

Lemma 3.3. *Let X and Y two real-valued random variables with marginal cdfs F_X and F_Y and bivariate copula C_η, depending on a copula parameter η. Let $\sigma_{X,Y}$, $\rho_{X,Y}$ and $\tau_{X,Y}$ denote (the population versions of) the covariance, Spearman's rank correlation coefficient and Kendall's tau, respectively, of X and Y. Then it holds:*

$$\sigma_{X,Y} = f_1(\eta) = \int_{\mathbb{R}^2} \left[C_\eta\{F_X(x), F_Y(y)\} - F_X(x)F_Y(y)\right] dx\, dy, \tag{3.4}$$

$$\rho_{X,Y} = f_2(\eta) = 12 \int_{[0,1]^2} C_\eta(u, v)\, du\, dv - 3, \tag{3.5}$$

$$\tau_{X,Y} = f_3(\eta) = 4 \int_{[0,1]^2} C_\eta(u, v)\, dC_\eta(u, v) - 1. \tag{3.6}$$

Proof. Equation (3.4) is due to Höffding (1940), Eq. (3.5) is Theorem 5.1.6. in Nelsen (2006) and (3.6) is Theorem 5.1.3 in Nelsen (2006). □

The "realized copula" method for empirical calibration of a one-dimensional parameter η of an m-variate copula essentially considers every of the $m(m-1)/2$ pairs of the m underlying random variables $X_1, \ldots, X_m$, inverts (3.4) each time with respect to η, replaces the population covariance by its empirical counterpart and aggregates the resulting $m(m-1)/2$ estimates in an appropriate way. More specifically, Fengler and Okhrin (2012) define for $1 \leq i < j \leq m$: $g_{ij}(\eta) = \hat{\sigma}_{ij} - f_1(\eta)$, set $\mathbf{g}(\eta) = (g_{ij}(\eta))_{1 \leq i < j \leq m}$, and propose to estimate

$$\hat{\eta} = \arg\min_{\eta} \mathbf{g}^\top(\eta)\mathbf{W}\mathbf{g}(\eta)$$

for an appropriate weight matrix $\mathbf{W} \in \mathbb{R}^{\binom{m}{2} \times \binom{m}{2}}$. In this, $\hat{\sigma}_{ij}$ denotes the empirical covariance of X_i and X_j. Indeed, any of the functions $f_\ell, \ell = 1, 2, 3$ corresponding to relationships (3.4)–(3.6) may be employed in this realized copula method. Moreover, they may be combined to estimate two- or three-dimensional copula parameters η.

In the particular context of estimating p-value copulae in multiple testing models, it is infeasible to actually draw independent replications of the vector $(p_i(X) : 1 \leq i \leq m)$ from the target population, because this would essentially mean to carry out the entire experiment several times. Hence, one typically employs resampling methods for estimating the dependency structure among the p-values, namely the parametric bootstrap or permutations if $H_1, \ldots, H_m$ correspond to marginal two-sample problems. Pollard and van der Laan (2004) compared both approaches and argued that the permutation method reproduces the correct null distribution only under some conditions. However, if these conditions are met, the permutation approach is often superior to bootstrapping (see also Westfall and Young (1993) and Meinshausen et al. (2011)). Furthermore, it is important to notice that both bootstrap and permutation-based methods estimate the joint distribution of $(p_i(X) : 1 \leq i \leq m)$ under the global null hypothesis H_0. Hence, the assumption that η is a nuisance parameter which does not depend on ϑ is an essential prerequisite for the applicability of such resampling methods for estimating η.

3.3 Closed Test Procedures

An important class of FWER-controlling multiple tests which do not exactly fall into one of the categories "margin-based" and "multivariate" is constituted by closed test procedures, introduced by Marcus et al. (1976).

Theorem 3.3. Let $\mathscr{H} = \{H_i : i \in I\}$ denote a $\cap$-closed system of hypotheses and $\varphi = (\varphi_i : i \in I)$ a coherent multiple test for $(\mathscr{X}, \mathscr{F}, \mathscr{P}, \mathscr{H})$ at local level α. Then, φ is a strongly FWER-controlling multiple test at FWER level α for $(\mathscr{X}, \mathscr{F}, \mathscr{P}, \mathscr{H})$.

Proof. Let $\vartheta \in \Theta$ with $I_0(\vartheta) \neq \emptyset$. Since $\mathscr{H}$ is $\cap$-closed, there exists an $i \in I$ with $H_i = \bigcap_{j \in I_0(\vartheta)} H_j$, and $\vartheta \in H_i$. Hence, for all $j \in I_0(\vartheta)$, we have $H_j \supseteq H_i$. Now, coherence of φ entails $\{\varphi_i = 1\} \supseteq \bigcup_{j \in I_0(\vartheta)} \{\varphi_j = 1\}$. We conclude that

$$\mathrm{FWER}_\vartheta(\varphi) = \mathbb{P}_\vartheta\left(\bigcup_{j \in I_0(\vartheta)} \{\varphi_j = 1\}\right) \leq \mathbb{P}_\vartheta(\{\varphi_i = 1\}) \leq \alpha,$$

because φ_i is a level α test. □

Notice that there is no restriction at all regarding the explicit form of the local level α tests φ_i in Theorem 3.3. One is completely free in choosing these tests. The decisive property of φ, however, is coherence. Not all multiple tests fulfill this property in the first place. This leads to the closed test principle, a "general solution to multiple testing problems" (Sonnemann (2008)).

Theorem 3.4 (Closure Principle, see Marcus et al. (1976), Sonnemann (2008)). Let $\mathscr{H} = \{H_i : i \in I\}$ denote a $\cap$-closed system of hypotheses and $\varphi = (\varphi_i : i \in I)$ an (arbitrary) multiple test for $(\mathscr{X}, \mathscr{F}, \mathscr{P}, \mathscr{H})$ at local level α. Then, we define the closed multiple test procedure (closed test) $\bar{\varphi} = (\bar{\varphi}_i : i \in I)$ based on φ by

$$\forall i \in I : \bar{\varphi}_i(x) = \min_{j: H_j \subseteq H_i} \varphi_j(x).$$

It holds:

(a) The closed test $\bar{\varphi}$ strongly controls the FWER at level α.
(b) For all $\emptyset \neq I' \subset I$, the "restricted" closed test $\bar{\varphi}' = (\bar{\varphi}_i : i \in I')$ is a strongly (at level α) FWER-controlling multiple test for $\mathscr{H}' = \{H_i : i \in I'\}$.
(c) Both tests $\bar{\varphi}$ and $\bar{\varphi}'$ are coherent.

Proof. The assertions follow immediately from the definitions of $\bar{\varphi}$ and $\bar{\varphi}'$ by making use of Theorem 3.3. □

Remark 3.1.

(a) The closed test $\bar{\varphi}$ based on φ rejects a particular hypothesis $H_i \in \mathscr{H}$ if and only if φ rejects H_i and all hypotheses $H_j \in \mathscr{H}$ of which H_i is a superset (implication).
(b) If $\mathscr{H}$ is not $\cap$-closed, then one can extend $\mathscr{H}$ by adding all missing intersection hypotheses, leading to the $\cap$-closed system of hypotheses $\bar{\mathscr{H}}$. If there are ℓ elementary hypotheses in $\mathscr{H}$, then $\bar{\mathscr{H}}$ can consist of up to $2^{\ell} - 1$ hypotheses. However, as we will demonstrate by specific examples, it is typically not necessary to test all elements in $\bar{\mathscr{H}}$ explicitly.
(c) Theorem 3.3 shows that under certain assumptions a multiple test at local level α is a strongly FWER-controlling multiple test at level α. Of course, the reverse statement is always true.
(d) If $\mathscr{H}$ is disjoint in the sense that $\forall i, j \in I, i \neq j : H_i \cap H_j = \emptyset$, and φ is a multiple test for $(\mathscr{X}, \mathscr{F}, \mathscr{P}, \mathscr{H})$ at local level α, then φ automatically strongly controls the FWER at level α, because φ is coherent and $\mathscr{H}$ is $\cap$-closed by the respective definitions. Often, there exist many possibilities for partitioning Θ in disjoint subsets, leading to the more general partitioning principle, see Finner and Strassburger (2002).
(e) If $I = \Theta$ and $H_\vartheta = \{\vartheta\}$ for all $\vartheta \in \Theta$, and if $\varphi = (\varphi_\vartheta : \vartheta \in \Theta)$ is a multiple test at local level α, then φ strongly controls the FWER at level α.

A nice application of the closed test principle is the problem of directional or type III errors, cf. Finner (1999) and references therein.

Example 3.3 (Two-sample t-test). Assume that we can observe $X = (X_{ij})$ for $i = 1, 2$ and $j = 1, \ldots, n_i$, that all X_{ij} are stochastically independent and

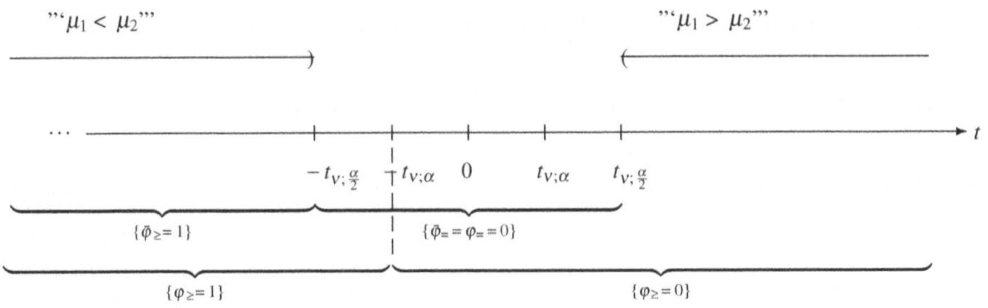

Fig. 3.3 Closed test for $\{H_=, H_\le, H_\ge\}$ in the two-sample Gaussian model

$X_{ij} \sim \mathcal{N}(\mu_i, \sigma^2)$ with unknown variance $\sigma^2 > 0$. Consider the hypothesis $H_= : \{\mu_1 = \mu_2\}$. The two-sample t-test $\varphi_=$ (say) for testing $H_=$ is based on the test statistic

$$T(X) = \sqrt{\frac{n_1 n_2}{n_1 + n_2}} \frac{\bar{X}_{1.} - \bar{X}_{2.}}{S}, \quad \text{where } S^2 = \frac{1}{\nu} \sum_{i=1}^{2} \sum_{j=1}^{n_i} (X_{ij} - \bar{X}_{i.})^2, \quad \nu = n_1 + n_2 - 2,$$

and is given by

$$\varphi_=(x) = \left\{ \begin{matrix} 1 & & > & \\ & |T(x)| & & t_{\nu;\alpha/2} \\ 0 & & \le & \end{matrix} \right\},$$

where $t_{\nu;\alpha/2}$ denotes the upper $\alpha/2$-quantile of Student's t-distribution with ν degrees of freedom. Let us restrict our attention to the case $\alpha \in (0, 1/2)$. The problem of directional or type III errors can be stated as follows. Assume that $H_=$ is rejected by $\varphi_=$. Can one then infer that $\mu_1 < \mu_2$ ($\mu_1 > \mu_2$) if $T(x) < -t_{\nu;\alpha/2}$ ($T(x) > t_{\nu;\alpha/2}$)? There is the possibility of an error of the third kind, namely, that $\mu_1 < \mu_2$ and $T(x) > t_{\nu;\alpha/2}$ ($\mu_1 > \mu_2$ and $T(x) < -t_{\nu;\alpha/2}$). The formal mathematical solution to this problem is given by the closed test principle. We add the two hypotheses $H_\le : \{\mu_1 \le \mu_2\}$ and $H_\ge : \{\mu_1 \ge \mu_2\}$ and notice that $H_= = H_\le \cap H_\ge$. Level α tests for $H_\le$ and $H_\ge$ are given by one-sided t-tests, say

$$\varphi_\le(x) = \left\{ \begin{matrix} 1 & & > & \\ & T(x) & & t_{\nu;\alpha} \\ 0 & & \le & \end{matrix} \right\}, \quad \varphi_\ge(x) = \left\{ \begin{matrix} 1 & & < & \\ & T(x) & & -t_{\nu;\alpha} \\ 0 & & \ge & \end{matrix} \right\}.$$

We construct the closed test $\bar{\varphi} = (\bar{\varphi}_\le, \bar{\varphi}_=, \bar{\varphi}_\ge)$, given by $\bar{\varphi}_= = \varphi_=$, $\bar{\varphi}_\le = \varphi_= \varphi_\le$, $\bar{\varphi}_\ge = \varphi_= \varphi_\ge$.

Due to the nestedness of the rejection regions of $\varphi_\le$ and $\bar{\varphi}_\le$ ($\varphi_\ge$ and $\bar{\varphi}_\ge$), see Fig. 3.3, it follows from Theorem 3.4 that type III errors are automatically controlled at level α, hence, one-sided decisions after two-sided testing are allowed in this model. The argumentation further shows that this is generally true for likelihood ratio test statistics, provided that the model implies an isotone likelihood ratio.

The presumably most intensively studied application of closed test procedures, however, is the context of analysis of variance models, where linear contrasts regarding the group-specific means are of interest. Since this field of application has already deeply been studied in earlier books (Hochberg and Tamhane (1987), Hsu (1996)), we abstain from covering it here. Closed test-related multiple testing strategies for systems of hypotheses with a tree structure have been worked out by Meinshausen (2008) and Goeman and Finos (2012); see also the references in these papers. In the latter case, power can be gained by exploiting the logical restrictions among the hypotheses which are given by the tree structure. This has some similarities to the methods considered by Westfall and Tobias (2007).

Acknowledgments The material on copula-based multiple tests originated from joint work with Taras Bodnar, Jakob Gierl and Jens Stange. Helmut Finner and Klaus Straßburger taught me everything I know about closed test procedures and the partitioning principle. Special thanks are due to Mareile Große Ruse for programming the LaTeX code used for the figures.

References

Arlot S, Blanchard G, Roquain E (2010) Some nonasymptotic results on resampling in high dimension. II: Multiple tests. Ann Stat 38(1):83–99. doi:10.1214/08-AOS668

Bauer P, Röhmel J, Maurer W, Hothorn L (1998) Testing strategies in multi-dose experiments including active control. Stat Med 17(18):2133–2146

Benjamini Y, Hochberg Y (1995) Controlling the false discovery rate: a practical and powerful approach to multiple testing. J Roy Stat Soc Ser B (Methodol) 57(1):289–300

Benjamini Y, Hochberg Y (2000) On the adaptive control of the false discovery rate in multiple testing with independent statistics. J Edu Behav Stat 25:60–83

Benjamini Y, Krieger AM, Yekutieli D (2006) Adaptive linear step-up procedures that control the false discovery rate. Biometrika 93(3):491–507

Bodnar T, Dickhaus T (2013) False Discovery Rate Control under Archimedean Copula. arXiv:1305.3897

Bonferroni CE (1935) Il calcolo delle assicurazioni su gruppi di teste. Studi in onore Salvatore Ortu Carboni 13–60.

Bonferroni CE (1936) Teoria statistica delle classi e calcolo delle probabilita. Pubbl. d. R. Ist. Super. di Sci. Econom. e Commerciali di Firenze 8. Firenze: Libr. Internaz. Seeber.

Bretz F, Hothorn T, Westfall P (2010) Multiple Comparisons Using R. Chapman and Hall/CRC

Dickhaus T (2013) Randomized p-values for multiple testing of composite null hypotheses. J Stat Plann Infer 143(11):1968–1979

Dickhaus T, Gierl J (2013) Simultaneous test procedures in terms of p-value copulae. Global Science and Technology Forum (GSTF). In: Proceedings on the 2nd Annual International Conference on Computational Mathematics, Computational Geometry and Statistics (CMCGS 2013), vol 2, pp 75–80

Dickhaus T, Stange J (2013) Multiple point hypothesis test problems and effective numbers of tests for control of the family-wise error rate. Calcutta Statist Assoc Bull, to appear

Dickhaus T, Strassburger K, Schunk D, Morcillo-Suarez C, Illig T, Navarro A (2012) How to analyze many contingency tables simultaneously in genetic association studies. Stat Appl Genet Mol Biol 11(4):Article 12

Dudoit S, van der Laan MJ (2008) Multiple testing procedures with applications to genomics. Springer Series in Statistics. Springer, New York

Farcomeni A (2009) Generalized augmentation to control false discovery exceedance in multiple testing. Scand J Stat 36(3):501–517

Fengler MR, Okhrin O (2012) Realized Copula. SFB 649 Discussion Paper 2012–034, Sonderforschungsbereich 649, Humboldt-Universität zu Berlin, Germany, available at http://sfb649.wiwi.hu-berlin.de/papers/pdf/SFB649DP2012-034.pdf

Finner H (1999) Stepwise multiple test procedures and control of directional errors. Ann Stat 27(1):274–289. doi:10.1214/aos/1018031111

Finner H, Gontscharuk V (2009) Controlling the familywise error rate with plug-in estimator for the proportion of true null hypotheses. J Roy Stat Soc B 71(5):1031–1048

Finner H, Strassburger K (2002) The partitioning principle: a powerful tool in multiple decision theory. Ann Stat 30(4):1194–1213

Finner H, Gontscharuk V, Dickhaus T (2012) False discovery rate control of step-up-down tests with special emphasis on the asymptotically optimal rejection curve. Scand J Stat 39:382–397

Goeman JJ, Finos L (2012) The inheritance procedure: multiple testing of tree-structured hypotheses. Stat Appl Genet Mol Biol 11(1):Article 11

Hochberg Y, Tamhane AC (1987) Multiple comparison procedures. Wiley Series in Probability and Mathematical Statistics. Applied Probability and Statistics. Wiley, New York

Hofert M, Mächler M, McNeil AJ (2012) Likelihood inference for Archimedean copulas in high dimensions under known margins. J Multivariate Anal 110:133–150. doi:10.1016/j.jmva.2012.02.019

Höffding W (1940) Maßstabinvariante Korrelationstheorie. Schr math Inst u Inst angew Math Univ Berlin 5:181–233

Holm SA (1979) A simple sequentially rejective multiple test procedure. Scand J Stat Theory Appl 6:65–70

Hothorn T, Bretz F, Westfall P (2008) Simultaneous inference in general parametric models. Biom J 50(3):346–363

Hsu JC (1996) Multiple comparisons: theory and methods. Chapman and Hall, London

van der Laan MJ, Dudoit S, Pollard KS (2004), Augmentation procedures for control of the generalized family-wise error rate and tail probabilities for the proportion of false positives. Stat Appl Genet Mol Biol 3: Article15

van der Laan MJ, Birkner MD, Hubbard AE (2005), Empirical Bayes and resampling based multiple testing procedure controlling tail probability of the proportion of false positives. Stat Appl Genet Mol Biol 4:Article29

Langaas M, Lindqvist BH, Ferkingstad E (2005) Estimating the proportion of true null hypotheses, with application to DNA microarray data. J Roy Stat Soc B 67(4):555–572. doi:10.1111/j.1467-9868.2005.00515.x

Lehmann EL, Romano JP (2005) Testing statistical hypotheses. Springer Texts in Statistics, 3rd edn. Springer, New York

Marcus R, Peritz E, Gabriel KR (1976) On closed test procedures with special reference to ordered analysis of variance. Biometrika 63(3):655–660

Meinshausen N (2008) Hierarchical testing of variable importance. Biometrika 95(2):265–278. doi:10.1093/biomet/asn007

Meinshausen N, Maathuis MH, Bühlmann P (2011) Asymptotic optimality of the Westfall-Young permutation procedure for multiple testing under dependence. Ann Stat 39(6):3369–3391. doi:10.1214/11-AOS946

Moskvina V, Schmidt KM (2008) On multiple-testing correction in genome-wide association studies. Genet Epidemiol 32:567–573

Nelsen RB (2006) An introduction to copulas. Springer series in statistics. 2nd edn. Springer,New York

Pollard KS, van der Laan MJ (2004) Choice of a null distribution in resampling-based multiple testing. J Stat Plann Infer 125(1–2):85–100. doi:10.1016/j.jspi.2003.07.019

Romano JP, Wolf M (2005a) Exact and approximate stepdown methods for multiple hypothesis testing. J Am Stat Assoc 100(469):94–108. doi:10.1198/016214504000000539

Romano JP, Wolf M (2005) Stepwise multiple testing as formalized data snooping. Econometrica 73(4):1237–1282. doi:10.1111/j.1468-0262.2005.00615.x

Romano JP, Shaikh AM, Wolf M (2008) Control of the false discovery rate under dependence using the bootstrap and subsampling. Test 17(3):417–442. doi:10.1007/s11749-008-0126-6

Scheffé H (1953) A method for judging all contrasts in the analysis of variance. Biometrika 40:87–110

Schweder T, Spjøtvoll E (1982) Plots of P-values to evaluate many tests simultaneously. Biometrika 69:493–502

Šidák Z (1967) Rectangular confidence regions for the means of multivariate normal distributions. J Am Stat Assoc 62:626–633. doi:10.2307/2283989

Sonnemann E (2008) General solutions to multiple testing problems. Translation of "Sonnemann, E. (1982). Allgemeine Lösungen multipler Testprobleme. EDV in Medizin und Biologie 13(4), 120–128". Biom J 50:641–656

Storey JD, Taylor JE, Siegmund D (2004) Strong control, conservative point estimation and simultaneous conservative consistency of false discovery rates: a unified approach. J. Roy. Stat. Soc. B, Stat Methodol 66(1):187–205

Tamhane AC, Liu W, Dunnett CW (1998) A generalized step-up-down multiple test procedure. Can J Stat 26(2):353–363. doi:10.2307/3315516

Troendle JF (1995) A stepwise resampling method of multiple hypothesis testing. J Am Stat Assoc 90(429):370–378. doi:10.2307/2291163

Troendle JF (2000) Stepwise normal theory multiple test procedures controlling the false discovery rate. J Stat Plann Infer 84(1–2):139–158. doi:10.1016/S0378-3758(99)00145-7

Van der Vaart A (1998) Asymptotic statistics. Cambridge Series in Statistical and Probabilistic Mathematics, 3. Cambridge University Press: Cambridge. doi:10.1017/CBO9780511802256

Westfall PH, Krishen A (2001) Optimally weighted, fixed sequence and gatekeeper multiple testing procedures. J Stat Plann Infer 99(1):25–40. doi:10.1016/S0378-3758(01)00077-5

Westfall PH, Tobias RD (2007) Multiple testing of general contrasts: truncated closure and the extended Shaffer-Royen method. J Am Stat Assoc 102(478):487–494. doi:10.1198/016214506000001338

Westfall PH, Young SS (1993) Resampling-based multiple testing: examples and methods for p-value adjustment. Wiley Series in Probability and Mathematical Statistics, Applied Probability and Statistics. Wiley, New York

Wiens BL (2003) A fixed sequence Bonferroni procedure for testing multiple endpoints. Pharmaceut Statist 2:211215

Yekutieli D, Benjamini Y (1999) Resampling-based false discovery rate controlling multiple test procedures for correlated test statistics. J Stat Plann Infer 82(1–2):171–196. doi:10.1016/S0378-3758(99)00041-5

Chapter 4
Simultaneous Test Procedures

Abstract We are considered with simultaneous test procedures (STPs) in the sense of Gabriel (1969). All marginal test statistics are compared with the same critical value, which is calculated under the global null hypothesis. We provide sufficient conditions for strong FWER control of STPs. Connections to multivariate analysis are drawn by introducing three families of multivariate distributions which play important roles in statistical models for multiple test problems. Models entailing a multivariate central limit theorem for least squares or maximum likelihood estimators, respectively, are investigated with respect to projection methods, leading to STPs. Probability bounds are employed to formalize the concept of effective numbers of tests. We discuss copula-based approaches to the construction of STPs and recall results from modern random field theory for constructing STPs by utilizing the topological structure of the underlying sample space.

At least since Gabriel (1969), a broad class of single-step multiple tests, so-called simultaneous test procedures (STPs), is established and systematically developed in the statistical literature.

Definition 4.1 (Simultaneous test procedure, Gabriel (1969)). Consider the (extended) multiple test problem $(\mathscr{X}, \mathscr{F}, (\mathbb{P}_\vartheta)_{\vartheta \in \Theta}, \mathscr{H}_{m+1})$ with $\mathscr{H}_{m+1} = \{H_i, i \in I^* := \{0, 1, \ldots, m\}\}$. Assume real-valued test statistics $T_i, i \in I^*$, which tend to larger values under alternatives. Then we call

(a) the pair $(\mathscr{H}_{m+1}, \mathscr{T})$ with $\mathscr{T} = \{T_i : i \in I^*\}$ a testing family.
(b) the multiple test $\varphi = (\varphi_i : i \in I^*)$ a simultaneous test procedure, if

$$\forall 0 \leq i \leq m : \varphi_i = \begin{cases} 1, & \text{if} \quad T_i > c_\alpha, \\ 0, & \text{if} \quad T_i \leq c_\alpha, \end{cases}$$

where the critical value c_α is determined such that $\forall \vartheta \in H_0 : \mathbb{P}_\vartheta(\{\varphi_0 = 1\}) = \mathbb{P}_\vartheta(\{T_0 > c_\alpha\}) \leq \alpha$.

T. Dickhaus, *Simultaneous Statistical Inference*,
DOI: 10.1007/978-3-642-45182-9_4,

Typical choices for T_0 are the sum-type statistic $T_0^{\text{sum}} = \sum_{i=1}^{m} a_i T_i$ for non-negative real constants a_i, and the max-statistic $T_0^{\max} = \max_{1 \le i \le m} T_i$. In this work, we will restrict our attention mainly to $T_0^{\max}$. Notice that $\{T_0^{\max} \le c_\alpha\} = \{\forall 1 \le i \le m : T_i \le c_\alpha\}$. Hence, for determining the critical value c_α with respect to FWER control, we get for an STP φ based on $T_0^{\max}$ and for $\vartheta^* \in H_0$ that

$$
\begin{aligned}
\text{FWER}_{\vartheta^*}(\varphi) &= \mathbb{P}_{\vartheta^*}(V_m > 0) = 1 - \mathbb{P}_{\vartheta^*}(V_m = 0) \\
&= 1 - \mathbb{P}_{\vartheta^*}(\forall 1 \le i \le m : T_i \le c_\alpha) \\
&= 1 - F_{(T_1,\ldots,T_m)}(c_\alpha, \ldots, c_\alpha) \qquad (4.1) \\
&= 1 - F_{T_0}(c_\alpha) = \mathbb{P}_{\vartheta^*}\left(T_0 > c_\alpha\right),
\end{aligned}
$$

where $F_{(T_1,\ldots,T_m)}$ and F_{T_0} denote the cdf. of $(T_1, \ldots, T_m)^\top$ and T_0, respectively, under ϑ^*. Equation (4.1) shows that c_α can equivalently be defined as an equi-coordinate $(1-\alpha)$-quantile of the joint distribution of the test statistics $(T_i : 1 \le i \le m)$ under the global hypothesis and that we can thus dispense with $T_0^{\max}$ in the sequel. Often, the latter joint distribution is unique, even if H_0 is composite. By these considerations, the theory of STPs is closely related to multivariate analysis. Therefore, we discuss three families of multivariate probability distributions which are important for practical applications in the next section.

Before doing so, we may briefly discuss justifications for calibrating c_α under the global hypothesis H_0. Of course, this yields weak FWER control of φ. However, often it also entails strong FWER control, namely, if the LFC for the STP φ is located in H_0. Several sufficient conditions for the latter have been discussed in the literature.

Definition 4.2.

(a) A testing family $(\mathcal{H}, \mathcal{T})$ is called monotone if for all $i, j \in I^*$ with $H_i \subseteq H_j$ and for almost all $x \in \mathcal{X}$, $T_i(x) \ge T_j(x)$.
(b) A testing family $(\mathcal{H}, \mathcal{T})$ is called joint if $\forall J \subseteq I : \forall \vartheta \in \bigcap_{j \in J} H_j$ the joint distribution of $\{T_j : j \in J\}$ is the same.
(c) A testing family $(\mathcal{H}, \mathcal{T})$ is called closed if $\mathcal{H}$ is closed under intersection.

Remark 4.1. Monotonicity in the sense of Definition 4.2.(a) is fulfilled if all T_i, $i \in I^*$, are likelihood ratio statistics.

Theorem 4.1 (Theorem 2 in Gabriel (1969)). *Assume that $(\mathcal{H}, \mathcal{T})$ is a monotone testing family and that $\varphi = (\varphi_i : i \in I^*)$ is an STP for the multiple test problem $(\mathcal{X}, \mathcal{F}, \mathcal{P}, \mathcal{H} = \{H_i, i \in I^*\})$ based on $(\mathcal{H}, \mathcal{T})$. If $(\mathcal{H}, \mathcal{T})$ is closed or joint, then φ controls the FWER strongly at level α.*

A slightly less restrictive distributional assumption than $(\mathcal{H}, \mathcal{T})$ being joint is given by the subset pivotality condition which has been introduced and extensively been made use of by Westfall and Young (1993) for resampling.

Definition 4.3 (Subset pivotality condition, cf. Westfall and Young (1993)). The vector $T = (T_1, \ldots, T_m)^\top$ is said to satisfy the subset pivotality condition (SPC), if

$$\forall \vartheta \in \Theta : \exists \vartheta^* \in H_0 : \mathbb{P}_{\vartheta}^{T_{I_0(\vartheta)}} = \mathbb{P}_{\vartheta^*}^{T_{I_0(\vartheta)}},$$

where the subvector $T_{I_0(\vartheta)}$ corresponds to the indices of true hypotheses in $\mathcal{H}$ under $\vartheta \in \Theta$.

Lemma 4.1. *Let* $(\mathcal{X}, \mathcal{F}, (\mathbb{P}_\vartheta)_{\vartheta \in \Theta}, \mathcal{H})$ *be a multiple test problem with a simple global hypothesis* $H_0 = \{\vartheta^*\}$ *(say). Let* $T = (T_1, \ldots, T_m)^\top$ *fulfill the SPC. Then,* ϑ^* *is the unique LFC for the STP* φ *induced by T, i. e.,*

$$\forall \vartheta \in \Theta : \mathit{FWER}_\vartheta(\varphi) \leq \mathit{FWER}_{\vartheta^*}(\varphi).$$

Consequently, φ *strongly controls the FWER at level* α*.*

Proof. Let $\vartheta \in \Theta$ be an arbitrary parameter value with resulting index set of true hypotheses $I_0 \equiv I_0(\vartheta)$. Let $O_{I_0} = \bigcap_{i \in I_0(\vartheta)} \{T_i \leq c_\alpha\}$ denote the event that none of the true null hypotheses is falsely rejected by the STP φ.

Utilizing the SPC, we obtain

$$\mathbb{P}_\vartheta(O_{I_0}) = \mathbb{P}_\vartheta\left(\bigcap_{i \in I_0(\vartheta)} \{T_i \leq c_\alpha\}\right) = \mathbb{P}_{\vartheta^*}\left(\bigcap_{i \in I_0(\vartheta)} \{T_i \leq c_\alpha\}\right) = \mathbb{P}_{\vartheta^*}(O_{I_0}),$$

and, consequently,

$$\mathrm{FWER}_\vartheta(\varphi) = 1 - \mathbb{P}_\vartheta(O_{I_0}) = 1 - \mathbb{P}_{\vartheta^*}(O_{I_0}) = \mathbb{P}_{\vartheta^*}\left(\bigcup_{i \in I_0(\vartheta)} \{\varphi_i = 1\}\right).$$

From the fact that $I_0(\vartheta) \subseteq I$, we conclude

$$\mathrm{FWER}_\vartheta(\varphi) \leq \mathbb{P}_{\vartheta^*}\left(\bigcup_{i=1}^{m} \{\varphi_i = 1\}\right) = \mathrm{FWER}_{\vartheta^*}(\varphi) \leq \alpha,$$

completing the proof. □

As outlined before, if Lemma 4.1 applies, the critical value c_α for the STP φ can be determined as an equi-coordinate $(1 - \alpha)$-quantile of the (joint) distribution $\mathbb{P}_{\vartheta^*}^T$ of $T = (T_1, \ldots, T_m)^\top$ under the global hypothesis. Uniqueness of ϑ^* can often be achieved by reparametrization. We will explain this by examples in Sect. 4.2.

4.1 Three Important Families of Multivariate Probability Distributions

As a matter of fact, simultaneous test procedures rely on results from multivariate analysis, in particular on distribution theory on $\mathbb{R}^m$. In this section, we briefly summarize properties of three families of multivariate probability distributions which are of importance for the multiple test problems that we investigate deeper in this work.

4.1.1 Multivariate Normal Distributions

A book-length treatment of multivariate normal distributions is provided by Tong (1990). Here, we only give a definition and the basic linearity properties. We refer to more specific properties at the respective occasions in the remainder of this work.

Definition 4.4 (Multivariate normal distribution). Let $X_1, \ldots, X_d$ iid. standard normal random variables on $\mathbb{R}$. Then we say that the random vector $X = (X_1, \ldots, X_d)^\top$ has a standard normal distribution on $\mathbb{R}^d$. Moreover, if $\Sigma = QQ^\top \in \mathbb{R}^{m \times m}$ denotes a positive definite, symmetric matrix with $Q \in \mathbb{R}^{m \times d}$ and we let $Y = QX + \mu,\ \mu \in \mathbb{R}^m$, then the random vector $Y = (Y_1, \ldots, Y_m)^\top$ possesses a (general) multivariate normal distribution on $\mathbb{R}^m$, and we write $Y \sim \mathcal{N}_m(\mu, \Sigma)$.

Theorem 4.2. *If $Y \sim \mathcal{N}_m(\mu, \Sigma)$, then Y has the density*

$$\phi_{\mu,\Sigma}(y) = (2\pi)^{-m/2} |\det \Sigma|^{-1/2} \exp\left(-\frac{1}{2}(y-\mu)^\top \Sigma^{-1}(y-\mu)\right)$$

with respect to the Lebesgue measure on $\mathbb{R}^m$. Furthermore, the first two moments of Y are given by

$$\forall 1 \leq j \leq m : \ \mathbb{E}[Y_j] = \mu_j, \quad \forall 1 \leq i,j \leq m : Cov\left(Y_i, Y_j\right) = \Sigma_{i,j}.$$

Proof. See Sect. 3.2 in Tong (1990). □

Theorem 4.3 (Affine transformations). *Let $Y \sim \mathcal{N}_m(\mu, \Sigma)$, $k \leq m$, $A \in \mathbb{R}^{k \times m}$ a matrix with maximum rank, and $b \in \mathbb{R}^k$. Then, the random vector $Z = AY + b$ has the k-dimensional normal distribution $\mathcal{N}_k(A\mu + b, A\Sigma A^\top)$.*

Proof. Theorem 3.3.3 in Tong (1990). □

4.1.2 Multivariate t-distributions

A comprehensive resource for multivariate t-distributions and their applications is the book by Kotz and Nadarajah (2004). Multivariate t-distributions are generalizations of Student's t-distribution on $\mathbb{R}$ to higher dimensions. In this, several possible generalizations exist. For our work, the version which considers a joint Studentization of an m-variate, normally distributed random vector is most important.

Definition 4.5. Let Y be an m-variate, centered random vector which is normally distributed with covariance matrix $\Sigma \in \mathbb{R}^{m \times m}$, where $\Sigma_{ii} \equiv \sigma^2 > 0$ for all $1 \leq i \leq m$. Let S with $\nu S^2/\sigma^2 \sim \chi^2_\nu$ for $\nu \geq 1$ be stochastically independent of Y. Then, letting R denote the correlation matrix corresponding to Σ, the distribution of

$$X = Y/S + \mu, \quad \mu \in \mathbb{R}^m, \tag{4.2}$$

is called a multivariate t-distribution with ν degrees of freedom, mean vector μ and correlation matrix R, and the pdf f_X of X is given by

$$f_X(x) = \frac{\Gamma((\nu+m)/2)}{(\pi\nu)^{m/2}\Gamma(\nu/2)|\det R|^{1/2}}\left[1 + \nu^{-1}(x-\mu)^\top R^{-1}(x-\mu)\right]^{-(\nu+m)/2}, \quad x \in \mathbb{R}^m.$$

4.1.3 Multivariate Chi-Square Distributions

Another extremely important family of probability distributions, which will be used at various places in the present work, is the family of multivariate chi-square distributions. Similarly to the situation for multivariate t-distributions, there exist several definitions of multivariate chi-square distributions in the literature, each of which has its respective origin in different models for real-life data. For our purposes, the following definition is most useful.

Definition 4.6 (Multivariate chi-square distribution). Let $m \geq 2$ and $\vec{\nu} = (\nu_1, \ldots, \nu_m)^\top$ a vector of positive integers. Let $(Z_{1,1}, \ldots, Z_{1,\nu_1}, Z_{2,1}, \ldots, Z_{2,\nu_2}, \ldots, Z_{m,1}, \ldots, Z_{m,\nu_m})$ denote $\sum_{k=1}^m \nu_k$ jointly normally distributed random variables with joint correlation matrix $R = (\rho(Z_{k_1,\ell_1}, Z_{k_2,\ell_2}) : 1 \leq k_1, k_2 \leq m, 1 \leq \ell_1 \leq \nu_{k_1}, 1 \leq \ell_2 \leq \nu_{k_2})$ such that for any $1 \leq k \leq m$ the random vector $\mathbf{Z}_k = (Z_{k,1}, \ldots, Z_{k,\nu_k})^\top$ has a standard normal distribution on $\mathbb{R}^{\nu_k}$. Let $\mathbf{Q} = (Q_1, \ldots, Q_m)^\top$, where

$$\forall 1 \leq k \leq m : Q_k = \sum_{\ell=1}^{\nu_k} Z^2_{k,\ell}. \tag{4.3}$$

Marginally, each Q_k is chi-square distributed with ν_k degrees of freedom. Hence, the distribution of the random vector $\mathbf{Q}$ is some multivariate (central) chi-square distribution with parameters m, $\vec{\nu}$ and R, and we write $\mathbf{Q} \sim \chi^2(m, \vec{\nu}, R)$.

Well-known special cases arise if all marginal degrees of freedom are identical, i. e., $\nu_1 = \nu_2 = \ldots = \nu_m \equiv \nu$ and the vectors $(Z_{1,1}, \ldots, Z_{m,1})^\top$, $(Z_{1,2}, \ldots, Z_{m,2})^\top$, ..., $(Z_{1,\nu}, \ldots, Z_{m,\nu})^\top$ are stochastically independent random vectors. If, in addition, the correlation matrices among the m components of these latter ν random vectors are all identical and equal to $\Sigma \in \mathbb{R}^{m \times m}$ (say), then the distribution of $\mathbf{Q}$ is that of the diagonal elements of a Wishart-distributed random matrix $S \sim W_m(\nu, \Sigma)$. This distribution is for instance given in Definition 3.5.7 of the textbook by Timm (2002). The case of potentially different correlation matrices $\Sigma_1, \ldots, \Sigma_\nu$ has been studied by Jensen (1970). From a practical perspective, it is remarkable that multivariate chi-square probabilities can exactly be computed, even for high dimensions, if the underlying correlation matrix R fulfills certain structural properties. We refer the reader to the article by Royen (2007) and references therein. Furthermore, the stochastic representation (4.3) allows to approximate multivariate chi-square probabilities by means of computer simulations up to (in principle) any precision.

The following lemma shows that among the components of a (generalized) multivariate chi-square distribution only non-negative correlations can occur.

Lemma 4.2. *Let* $\mathbf{Q} \sim \chi^2(m, \vec{\nu}, R)$. *Then, for any pair of indices* $1 \leq k_1, k_2 \leq m$, *it holds*

$$0 \leq Cov(Q_{k_1}, Q_{k_2}) \leq 2\sqrt{\nu_{k_1} \nu_{k_2}}. \tag{4.4}$$

Proof. Without loss of generality, assume $k_1 = 1$ and $k_2 = 2$. Simple probabilistic calculus now yields

$$\begin{aligned} \text{Cov}(Q_1, Q_2) &= \text{Cov}\left(\sum_{i=1}^{\nu_1} Z_{1,i}^2, \sum_{j=1}^{\nu_2} Z_{2,j}^2\right) \\ &= \sum_{i=1}^{\nu_1}\sum_{j=1}^{\nu_2} \text{Cov}(Z_{1,i}^2, Z_{2,j}^2) = 2\sum_{i=1}^{\nu_1}\sum_{j=1}^{\nu_2} \rho^2(Z_{1,i}, Z_{2,j}) \geq 0. \end{aligned}$$

The upper bound in (4.4) follows directly from the Cauchy-Schwarz inequality, because the variance of a chi-square distributed random variable with ν degrees of freedom equals 2ν. □

4.2 Projection Methods Under Asymptotic Normality

For a broad class of statistical models which are relevant for practical applications, the maximum likelihood estimator for the vector of unknown model parameters is (at least asymptotically for large sample sizes) unbiased and normally distributed. Let us discuss a few examples.

Definition 4.7 (Multiple linear regression model). Consider the sample space $(\mathbb{R}^n, \mathscr{B}(\mathbb{R}^n))$ and real-valued, stochastically independent observables $Y_1, \ldots, Y_n$ such that

$$\forall 1 \leq i \leq n: \quad Y_i = f(x_{i,1}, \ldots, x_{i,k}) + \varepsilon_i = \sum_{j=1}^{k} \vartheta_j x_{i,j} + \varepsilon_i. \tag{4.5}$$

In this, $X = (x_{ij})_{1 \leq i \leq n, 1 \leq j \leq k}$ is a deterministic $(n \times k)$ matrix, where $k \leq n$, called the design matrix. The vector $\vartheta = (\vartheta_1, \ldots, \vartheta_k)^\top \in \mathbb{R}^k$ is the parameter of interest. Abbreviating $Y = (Y_1, \ldots, Y_n)^\top \in \mathbb{R}^n$ (the response vector) and $\varepsilon = (\varepsilon_1, \ldots, \varepsilon_n)^\top \in \mathbb{R}^n$ (the vector of error terms), we obtain the matrix representation of (4.5), given by

$$Y = X\vartheta + \varepsilon. \tag{4.6}$$

We make the following additional assumptions.

(a) The design matrix has maximum rank, such that $X^\top X \in \mathbb{R}^{k \times k}$ is positive definite and invertible.
(b) The error terms are iid. with $\mathbb{E}[\varepsilon_1] = 0$ and $0 < \sigma^2 = \mathrm{Var}(\varepsilon_1) < \infty$.

Notice that Definition 4.7 covers analysis of variance models by choosing X appropriately (containing group membership indicators). The components ϑ_j, $1 \leq j \leq k$, then correspond to group-specific means. Hence, k-sample problems with localized comparisons can be modeled with (4.6).

Theorem 4.4. *Under the multiple linear regression model from Definition 4.7, the least squares estimator of* $\vartheta \in \mathbb{R}^k$ *is given by*

$$\hat{\vartheta} = (X^\top X)^{-1} X^\top Y.$$

Using (4.6)*, this leads to*

$$\hat{\vartheta} - \vartheta = (X^\top X)^{-1} X^\top \varepsilon.$$

If $\varepsilon \sim \mathscr{N}_n(0, \sigma^2 I_n)$*, then* $\hat{\vartheta}$ *is the maximum likelihood estimator of* ϑ *and it holds* $\hat{\vartheta} \sim \mathscr{N}_k(\vartheta, \sigma^2 (X^\top X)^{-1})$*. In the general case, assume that the following two conditions hold true.*

(i) $n^{-\frac{1}{2}} \max_{1 \leq i \leq n, 1 \leq j \leq k} |x_{i,j}| \to 0$, $n \to \infty$.
(ii) $n^{-1} X_n^\top X_n \to V$, $V \in \mathbb{R}^{k \times k}$ *symmetric and positive definite.*

Then, for $n \to \infty$*,*

$$\mathscr{L}\left(\sqrt{n}\,\{\hat{\vartheta}(n) - \vartheta\}\right) \xrightarrow{w} \mathscr{N}_k\left(0, \sigma^2 V^{-1}\right).$$

Definition 4.8 (Generalized linear model (GLM)). Let $(\mathcal{Y}^n, \mathcal{F}^n, \bigotimes_{i=1}^n \mathbb{P}_{\xi_i})$ denote a product model with unknown parameters $\xi_i \in \Xi \subseteq \mathbb{R}$ for all $1 \leq i \leq n$. Denote the stochastically independent observables by $Y_1, \ldots, Y_n$, where each Y_i takes values in $\mathcal{Y} \subseteq \mathbb{R}$. Then, $(\mathcal{Y}^n, \mathcal{F}^n, \bigotimes_{i=1}^n \mathbb{P}_{\xi_i})$ is called a generalized linear model (GLM) if the following three conditions hold true.

(i) For any $1 \leq i \leq n$, $\mathbb{P}_{\xi_i}$ is an element of an exponential family with likelihood function of the form

$$L(\xi_i, y_i) = a(\xi_i)b(y_i)\exp(y_i \cdot T(\xi_i)).$$

The quantity $T(\xi)$ is called the natural parameter of the exponential family, $\xi \in \Xi$.

(ii) A design matrix $X \in \mathbb{R}^{n \times k}$ is given, leading to the vector $\eta = (\eta_1, \ldots, \eta_n)^\top$ of linear predictors, where $\eta_i = \sum_{j=1}^k \vartheta_j x_{ij}$, $1 \leq i \leq n$, for (unknown) coefficients $\vartheta_1, \ldots, \vartheta_k$, which are the parameters of interest. In matrix form, we have $\eta = X\vartheta$, $\vartheta = (\vartheta_1, \ldots, \vartheta_k)^\top$.

(iii) Let $\mu_i = \mathbb{E}[Y_i|X_i = x_i]$ denote the (conditional) expected value of Y_i given $X_i = x_i = (x_{i1}, \ldots, x_{ik})$. It exists a link function g such that

$$\forall 1 \leq i \leq n : \eta_i = g(\mu_i) \Leftrightarrow g(\mu_i) = \sum_{j=1}^k \vartheta_j x_{ij}.$$

The canonical link maps μ_i onto the natural parameter of the exponential family, i. e.,

$$g(\mu_i) = T(\xi_i) \Leftrightarrow T(\xi_i) = \sum_{j=1}^k \vartheta_j x_{ij}.$$

In general, the maximum likelihood estimator (MLE) $\hat{\vartheta}$ of the parameter vector $\vartheta = (\vartheta_1, \ldots, \vartheta_k)^\top$ of a GLM cannot be written in closed form. However, existence and uniqueness of $\hat{\vartheta}$ are guaranteed and efficient numerical algorithms exist for its computation. The following theorem establishes the limiting distribution of $\hat{\vartheta}$ when the sample size n tends to infinity.

Theorem 4.5 (Multivariate central limit theorem). *Let $\hat{\vartheta}(n)$ denote the MLE of the parameter vector $\vartheta = (\vartheta_1, \ldots, \vartheta_k)^\top$ of a GLM with canonical link, depending on the sample size n. If all k covariates (corresponding to the columns of X) have compact support and if $(X_n^\top X_n)^{-1} \to 0$, $n \to \infty$, then it holds*

$$\hat{\vartheta}(n) \underset{asympt.}{\sim} \mathcal{N}_k(\vartheta, F_n^{-1}(\vartheta)), \quad where \ F_n(\vartheta) = X_n^\top Cov_n(Y) X_n.$$

The result remains to hold true if $F_n(\vartheta)$ is replaced by $F_n(\hat{\vartheta}(n))$.

Proof. Satz 2.2 in Chap. 7 of Fahrmeir and Hamerle (1984).

Remark 4.2. Since the response variables are stochastically independent, it holds $\mathrm{Cov}_n(Y) = \mathrm{diag}\left(\mathrm{Var}(Y_i|\vec{X}_i = \vec{x}_i) : 1 \leq i \leq n\right)$. These (conditional) variances depend on ϑ (via ξ_i which is modeled as a function of ϑ).

By virtue of Theorem 4.3, Theorems 4.4 and 4.5 also imply the (asymptotic) distribution of the vector $T = C\hat{\vartheta} - d$ of test statistics for the system $\mathscr{H}_m$ of linear hypotheses

$$C\vartheta = d, \tag{4.7}$$

where $C \in \mathbb{R}^{m \times k}$ is called a contrast matrix and $d \in \mathbb{R}^m$ a given vector. Notice that we interpret (4.7) here as a system of m hypotheses, meaning that we define H_i by line i of the system of equations in (4.7), where $1 \leq i \leq m$. Each H_i encodes one linear restriction concerning (components of) ϑ. In contrast, the classical F-test in multiple linear regression analysis considers (4.7) as one single hypothesis, namely, the global hypothesis $H_0 = \bigcap_{i=1}^m H_i$. Although each H_i and even H_0 may be a composite hypothesis (with respect to ϑ), the (asymptotic) joint distribution of T is under any $\vartheta \in H_0$ a centered m-variate normal distribution the covariance matrix of which depends on the model and on the contrast matrix C. If a linear model with normally distributed error terms is assumed, then the exact distribution of a Studentized version of T under H_0 is given by a multivariate t-distribution, cf. Section 4.1.2. Hence, in models for which Theorem 4.4 or Theorem 4.5 applies, STPs for systems of linear hypotheses at (asymptotic) FWER level α can be derived by computing equi-coordinate $(1 - \alpha)$-quantiles of multivariate normal or multivariate t-distributions. This is leads to the projection tests derived by Hothorn et al. (2008). Denoting the two-sided equi-coordinate $(1 - \alpha)$-quantile of the (asymptotic) joint distribution of T under H_0 by c_α, H_i gets rejected if $|T_i|$ exceeds c_α. Figure 4.1 displays the situation for $m = 2$, standard normal marginal distributions of T_1 and T_2, and three different values of the correlation $\rho(T_1, T_2)$. One observes that with growing correlation the necessary adjustment for multiplicity (i. e., the value c_α) decreases in comparison to the independent case (which corresponds to a Šidák correction).

Remark 4.3.

(a) If hypotheses shall be weighted for importance, one can consider a rectangle instead of a square in Fig. 4.1.
(b) In cases where Theorem 4.5 applies, the limiting covariance matrix of $\hat{\vartheta}$ and, hence, that of T may depend on ϑ, cf. Remark 4.2. In such cases, one will in practice apply Theorem 4.5 with the estimate $F_n(\hat{\vartheta}(n))$ of $F_n(\vartheta)$. In the strictest sense, the resulting multiple test is not an STP according to Definition 4.1, because c_α is not calibrated under the assumption that the (limiting) covariance matrix of $\hat{\vartheta}$ is as under $\vartheta^* \in H_0$.

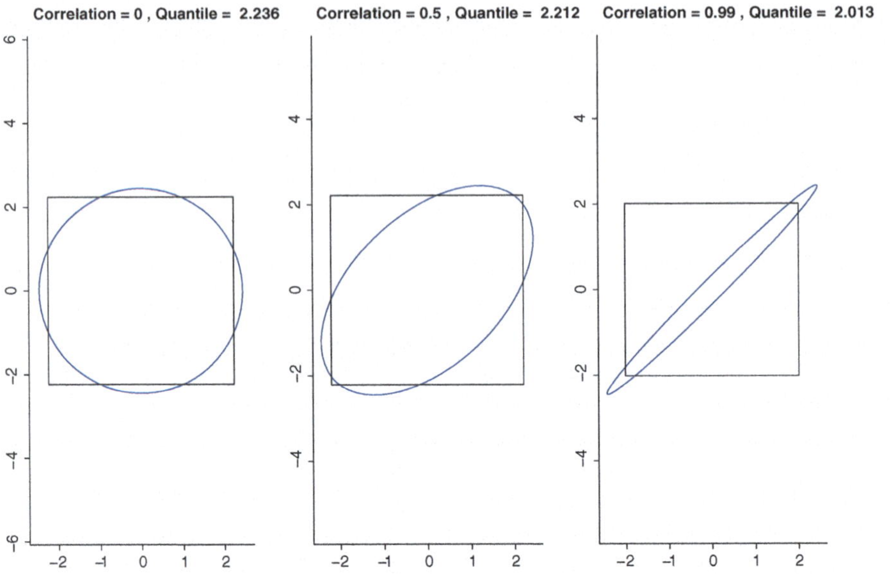

Fig. 4.1 Graphical illustration of two-sided equi-coordinate 95 %-quantiles for a bivariate normal distribution with standard normal margins and correlation coefficient ρ, for $\rho = 0$ (*left panel*) $\rho = 0.5$ (*middle panel*), and $\rho = 0.99$ (*right panel*). Both the respectively displayed square and ellipse contain 95 % of the distributional mass of the respective bivariate normal distribution

(c) Multivariate central limit theorems for maximum likelihood estimators can be established for further parametric model classes, too.

Applications of the projection methods under (asymptotic) normality discussed in this section to the field of genetics have been exemplified by Conneely and Boehnke (2007).

4.3 Probability Bounds and Effective Numbers of Tests

If m is large, it will often be infeasible to work with the full joint distribution of T under H_0, even if it is unique and exactly known. For example, the `R`-package `mvtnorm` computes multivariate t- and normal probabilities up to dimension 1000, but not for higher dimensions. In cases where $\mathbb{P}^T_{\vartheta^*}$ is intractable, one often works with conservative approximations of the probability $\mathbb{P}_{\vartheta^*}(\forall 1 \leq i \leq m : T_i \leq c_\alpha)$ which has to be computed in order to determine c_α, leading to the theory of probability bounds.

4.3.1 Sum-Type Probability Bounds

Assume that the (or: an) LFC ϑ^* for an STP φ is located in the global hypothesis H_0. Then, for any $\vartheta \in \Theta$, the FWER of φ under ϑ is upper-bounded by

$$\text{FWER}_{\vartheta^*}(\varphi) = \mathbb{P}_{\vartheta^*}\left(\bigcup_{i=1}^{m} A_i\right),$$

where $A_i = \{T_i > c_\alpha\}$. Now suppose you can find an upper bound $b(\mathbb{P}_{\vartheta^*}, \alpha_{\text{loc.}})$ such that

$$\mathbb{P}_{\vartheta^*}\left(\bigcup_{i=1}^{m} A_i\right) \leq b(\mathbb{P}_{\vartheta^*}, \alpha_{\text{loc.}}), \tag{4.8}$$

where $\alpha_{\text{loc.}}$ has the interpretation of a multiplicity-adjusted local significance level to be used in every marginal test problem H_i versus K_i, $1 \leq i \leq m$, as explained in Sect. 2.2.5. Then, φ can (conservatively) be calibrated for strong FWER control at level α by solving the equation $b(\mathbb{P}_{\vartheta^*}, \alpha_{\text{loc.}}) = \alpha$ for $\alpha_{\text{loc.}}$. This connects the theory of STPs with that of probability bounds. Determining bounds $b(\mathbb{P})$ such that

$$\mathbb{P}\left(\bigcup_{i=1}^{m} A_i\right) \leq b(\mathbb{P}) \tag{4.9}$$

for an arbitrary probability measure $\mathbb{P}$ and $\mathbb{P}$-measurable events $(A_i)_{1 \leq i \leq m}$ is a classical topic in probability theory, see, e. g., Rényi (1961), Galambos and Rényi (1968), Hunter (1976), Worsley (1982), Efron (1997), Ninomiya and Fujisawa (2007), and Naiman and Wynn (1992, 1997, 2001, 2005).

The union structure of the event $\bigcup_{i=1}^{m} A_i$ suggests bounds of sum-type. The certainly most basic of such sum-type probability bounds is the Bonferroni bound, i. e., $b(\mathbb{P}) = \sum_{i=1}^{m} \mathbb{P}(A_i)$, leading to the Bonferroni correction $\alpha_{\text{loc.}} = \alpha/m$ as explained in Example 3.1. Since there exist examples where this bound yields equality in (4.9), an improved bound can only be derived if the joint distribution $\mathbb{P}^{T}_{\vartheta^*}$ of test statistics under ϑ^* possesses certain properties. In particular, the dependency structure among $T_1, \ldots, T_m$ under ϑ^* crucially matters. Hence, we discuss some improved sum-type probability bounds for dependent test statistics in the remainder of this section.

Theorem 4.6 (Corollary 1 by Worsley (1982)). *Let $(\mathcal{X}, \mathcal{F}, \mathbb{P})$ be a probability space and $(A_i)_{1 \leq i \leq m}$ a finite sequence of $\mathbb{P}$-measurable events. Then it holds*

$$\mathbb{P}\left(\bigcup_{i=1}^{m} A_i\right) \leq \sum_{i=1}^{m} \mathbb{P}(A_i) - \sum_{i=1}^{m-1} \mathbb{P}(A_i \cap A_{i+1}). \tag{4.10}$$

Application of the bound on the right-hand side of (4.10) to events of the form $A_i = \{T_i > c_\alpha\}$ exploits the bivariate marginal distributions of the random variables $T_1, \ldots, T_m$, in contrast to the Bonferroni bound which only considers the m univariate marginal distributions. In the special case that $T = (T_1, \ldots, T_m)^\top$ has an (asymptotic) normal distribution, bivariate distributions are characterized by the correlations coefficients between the components in T and the "length heuristic" considered by Efron (1997) is a useful tool, at least for positive pairwise correlations.

Theorem 4.7 (*W* **formula by Efron (1997)).** *Let $T = (T_1, \ldots, T_m)^\top$ denote a vector of (correlated) standard normal random variables. For $2 \leq j \leq m$, let ρ_j denote the correlation coefficient of T_{j-1} and T_j. Furthermore, let $c = c(\alpha_{loc.})$ for $\alpha_{loc.} \in (0, 1)$ denote the upper $\alpha_{loc.}$-quantile of the standard normal law on $\mathbb{R}$. Then it holds*

$$\mathbb{P}\left(\bigcup_{j=1}^{m}\{T_j > c\}\right) \leq \alpha_{loc.} + \phi(c)\sum_{j=2}^{m}\frac{2\Phi(cL_j/2) - 1}{c}, \tag{4.11}$$

where $L_j = \arccos(\rho_j)$.

Theorem 4.7 may be used in cases where Theorem 4.4 or Theorem 4.5 applies, but the dimensionality m prohibits calculation of the full joint distribution of T.

4.3.2 Product-Type Probability Bounds

As discussed around (4.1), the FWER of an STP φ with LFC ϑ^* located in H_0 has the maximum

$$\mathrm{FWER}_{\vartheta^*}(\varphi) = 1 - \mathbb{P}_{\vartheta^*}\left(\bigcap_{i=1}^{m}\{T_i \leq c_\alpha\}\right)$$

and hence,

$$\mathrm{FWER}_{\vartheta^*}(\varphi) \leq \alpha \iff \mathbb{P}_{\vartheta^*}\left(\bigcap_{i=1}^{m}\{T_i \leq c_\alpha\}\right) \geq 1 - \alpha. \tag{4.12}$$

Somewhat conversely to the methods discussed in Sect. 4.3.1, we can therefore conservatively calibrate φ by finding a bound $\beta(\mathbb{P}_{\vartheta^*}, \alpha_{\mathrm{loc.}})$ such that

$$\mathbb{P}_{\vartheta^*}\left(\bigcap_{i=1}^{m}\{T_i \leq c_\alpha\}\right) \geq \beta(\mathbb{P}_{\vartheta^*}, \alpha_{\mathrm{loc.}}) \tag{4.13}$$

and solving the equation $\beta(\mathbb{P}_{\vartheta^*}, \alpha_{\mathrm{loc.}}) = 1 - \alpha$ for $\alpha_{\mathrm{loc.}}$. Here, the intersection structure of the event $\bigcap_{i=1}^{m}\{T_i \leq c_\alpha\}$ suggests bounds of product-type. It turns out that such bounds can naturally be derived if the joint distribution $\mathbb{P}^T_{\vartheta^*}$ possesses positive dependency properties of certain kinds. If $T_1, \ldots, T_m$ are jointly independent under ϑ^*, then, for any cutoff $c_\alpha \in \mathbb{R}$, the Šidák bound $\beta(\mathbb{P}_{\vartheta^*}, \alpha_{\text{loc.}}) = \prod_{i=1}^m \mathbb{P}_{\vartheta^*}(T_i \le c_\alpha)$ is exact and therefore valid for calibrating φ. This bound remains valid under the positive dependency concepts in the following definition.

Definition 4.9 (Concepts of positive dependence). Let $(\mathscr{X}, \mathscr{F}, \mathbb{P})$ be a probability space and let $T = (T_1, \ldots, T_m)^\top : \mathscr{X} \to S^m$ be a random vector, where $(S, \mathscr{S})$ is a measurable space with S a subset of $\mathbb{R}$. In all definitions below, $t = (t_1, \ldots, t_m)^\top$ denotes an arbitrary element of S^m.

(i) For $1 \le j \le m$, let $P_j \equiv P_j(t) = \mathbb{P}\left(\bigcap_{h=1}^j \{T_h \le t_h\}\right)$, $\gamma_{j,1} \equiv \gamma_{j,1}(t) = \mathbb{P}(T_j \le t_j)$, and

$$\gamma_{j,i} \equiv \gamma_{j,i}(t) = \mathbb{P}\left(T_j \le t_j \mid \bigcap_{h=j-i+1}^{j-1} \{T_h \le t_h\}\right), \ 1 < i \le j.$$

Due to chain factorization, it holds $P_m = P_i \cdot \prod_{j=i+1}^m \gamma_{j,j}$ for every fixed $1 \le i \le m-1$. Following Block et al. (1992), we call

$$\beta_i = P_i \cdot \prod_{j=i+1}^{m} \gamma_{j,i} \tag{4.14}$$

the product-type probability bound of order i. Moreover, we call T sub-Markovian of order i (SM$_i$), if $\gamma_{k,k} \ge \gamma_{k,i}$ for all $i+1 \le k \le m$, entailing that $P_m \ge \beta_i$. We call T monotonically sub-Markovian of order i (MSM$_i$), if $\gamma_{k,k} \ge \gamma_{k,i} \ge \gamma_{k,i-1} \ge \cdots \ge \gamma_{k,1}$ for $k \ge i$ and $\gamma_{k,k} \ge \gamma_{k,k-1} \ge \cdots \ge \gamma_{k,1}$ for $i > k \ge 1$, entailing $P_m \ge \beta_i \ge \beta_{i-1} \ge \cdots \ge \beta_1$.

(ii) T is called positive lower orthant dependent (PLOD), if

$$\mathbb{P}(T_1 \le t_1, \ldots, T_m \le t_m) \ge \prod_{j=1}^m \mathbb{P}(T_j \le t_j).$$

In other words, PLOD is equivalent to $P_m \ge \beta_1$.

(iii) T is called multivariate totally positive of order 2 (MTP$_2$), if its distribution $\mathbb{P}^T$ on $(S^m, \mathscr{S}^{\otimes m})$ has a probability density function $f : S^m \to [0, \infty)$ with respect to a measure $\sigma^{\otimes m}$, such that for all $u, v \in S^m$:

$$f(u) \cdot f(v) \le f(\min(u, v)) \cdot f(\max(u, v)),$$

where the minimum or maximum, respectively, is being taken component-wise.

Part (i) of the latter definition shows the usefulness of the MSM_i property for deriving product-type probability bounds. The following proposition shows that there exists a hierarchy in the concepts of positive dependence introduced in Definition 4.9.

Proposition 4.1. *Under the assumptions of Definition 4.9, it holds*

(i) MTP_2 *implies* MSM_{m-1}.
(ii) MSM_i *implies* MSM_h *for all* $1 \leq h \leq i$. *In particular,* MSM_i *for* $i \geq 2$ *implies PLOD.*

Proof. The assertions under (ii) are obvious and the assertion under (i) has been proven by Glaz and Johnson (1984). □

Combining Proposition 4.1 and the reasoning in part (i) of Definition 4.9, we obtain that nested product-type probability bounds $P_m \equiv \beta_m \geq \beta_i \geq \beta_{i-1} \geq \cdots \geq \beta_1$ for calibrating the STP φ can immediately be derived in terms of the i-variate marginal distributions of $T = (T_1, \ldots, T_m)^\top$ whenever $\mathbb{P}_{\vartheta^*}$ is MTP_2. Furthermore, some characterizations of MTP_2 and MSM_i exist for the important families of multivariate distributions that we have introduced in Sect. 4.1.

Proposition 4.2.

(a) Let $X = (X_1, \ldots, X_m)$ *denote a centered multivariate Gaussian random vector,* $X \sim \mathcal{N}_m(0, \Sigma)$, *with* Σ *positive definite and let* $|X| = (|X_1|, \ldots, |X_m|)$.

(i) If all entries of Σ *are non-negative, then* X *is* MTP_2.
(ii) Independently of Σ, $|X|$ *is PLOD.*
(iii) Independently of Σ, $\beta_2 \geq \beta_1$ *for* $T = |X|$.
(iv) $|X|$ *is* MTP_2 *if and only if there exists a diagonal matrix* D *with diagonal elements* ± 1 *such that the off-diagonal elements of* $-D\Sigma^{-1}D$ *are all non-negative.*

(b) Assume that $X = (X_1, \ldots, X_m)$ *has a centered multivariate t-distribution with* ν *degrees of freedom and correlation matrix* $R = (\rho_{ij})_{1 \leq i,j \leq m}$.

(i) If $\rho_{ij} \geq 0$ *for all* $i \neq j$, *then* X *is PLOD.*
(ii) Independently of R, $|X|$ *is PLOD, where* $|X|$ *is defined in analogy to part (a).*
(iii) Independently of R, $\beta_2 \geq \beta_1$ *for* $T = |X|$.

(c) Let $T = (T_1, \ldots, T_m)$ *follow a multivariate central chi-square distribution with* ν *degrees of freedom in every marginal.*

(i) If the distribution of T *is as in Definition 3.5.7 in Timm (2002), and all diagonal elements of* Σ *are equal to* 1, *then, independently of the off-diagonal elements of* Σ, T *is PLOD.*
(ii) For T *as in part (i), it holds* $\beta_2 \geq \beta_1$.
(iii) Under exchangeability (entailing equi-correlation) in the sense that each T_j *can be represented as* $T_j = Z_j + Z_0$ *for stochastically independent, chi-square distributed variates* $Z_0, Z_1, \ldots, Z_m$, T *is* MTP_2.

Proof. Part (a).(i) follows from Sect. 4.3.3 in Tong (1990). Part (a).(ii) is Corollary 1 in Šidák (1967). To prove part (a).(iii), we first notice that all bivariate marginal distributions of an m-variate normal distribution are bivariate normal, $m \geq 2$. Since the PLOD property for the absolute values of a Gaussian random vector is valid without any assumptions on the dimension or on Σ, we have that every pair $(|X_k|, |X_\ell|)$ is PLOD. This entails $\beta_2 \geq \beta_1$. Part (a).(iv) is Theorem 3.1 in Karlin and Rinott (1980). Part (b).(i) is Corollary 9.2.2 and part (b).(ii) is Corollary 9.2.3 in Tong (1990). Part (b).(iii) can be proved in analogy to part (a).(iii). To prove part (c).(i), we notice that the distribution of T is equal to the joint distribution of the diagonal elements $S_{1,1}, \ldots, S_{m,m}$ of a Wishart-distributed random matrix $S \sim W_m(\nu, \Sigma)$. Corollary 4.1 in Das Gupta et al. (1972) yields the assertion. Part (c).(ii) can again be proved in analogy to part (a).(iii). Finally, part (c).(iii) is a consequence of Example 3.5. in Karlin and Rinott (1980). □

4.3.3 *Effective Numbers of Tests*

An interesting technique that has received much attention in the context of multiple testing for genetic applications is to transform probability bounds into "effective numbers of tests". We define the effective number of tests corresponding to a sum-type (product-type) probability bound as follows.

Definition 4.10. Let $(\mathcal{X}, \mathcal{F}, (\mathbb{P}_\vartheta)_{\vartheta \in \Theta}, \mathcal{H}_{m+1})$ denote an (extended) multiple test problem in the sense of Definition 4.1 and $\varphi = (\varphi_i : 1 \leq i \leq m)$ an STP for $(\mathcal{X}, \mathcal{F}, (\mathbb{P}_\vartheta)_{\vartheta \in \Theta}, \mathcal{H}_{m+1})$. Furthermore, let an FWER level α be given.

(i) Assume that φ can (conservatively) be calibrated by a sum-type probability bound $b(\mathbb{P}_{\vartheta^*}, \alpha_{\text{loc.}})$. Then, the effective number of tests $M_{\text{eff.}} \equiv M_{\text{eff.}}(\alpha)$ is defined as the unique solution of the equation

$$M_{\text{eff.}} \alpha_{\text{loc.}} = b(\mathbb{P}_{\vartheta^*}, \alpha_{\text{loc.}}) = \alpha. \tag{4.15}$$

(ii) Assume that φ can (conservatively) be calibrated by a product-type probability bound $\beta(\mathbb{P}_{\vartheta^*}, \alpha_{\text{loc.}})$. Then, the effective number of tests $M_{\text{eff.}} \equiv M_{\text{eff.}}(\alpha)$ is defined as the unique solution of the equation

$$(1 - \alpha_{\text{loc.}})^{M_{\text{eff.}}} = \beta(\mathbb{P}_{\vartheta^*}, \alpha_{\text{loc.}}) = 1 - \alpha. \tag{4.16}$$

Example 4.1 (Effective numbers of tests).

(i) For the Bonferroni bound $b(\mathbb{P}_{\vartheta^*}, \alpha_{\text{loc.}})$ as well as for the Šidák bound $\beta(\mathbb{P}_{\vartheta^*}, \alpha_{\text{loc.}})$ it holds $M_{\text{eff.}} = m$.
(ii) Let $T = (T_1, \ldots, T_m)$ be such that the product-type probability bound β_2 defined in (4.14) applies. Define cut-offs $c = (c_1, \ldots, c_m) \in \mathbb{R}^m$ such that $\forall j \in \{1, \ldots, m\} : \mathbb{P}_{\vartheta_j^*}(\varphi = 1) = \mathbb{P}_{\vartheta_j^*}(T_j > c_j) = \alpha_{\text{loc.}}$ for a local significance

level $\alpha_{\text{loc.}} \in (0, 1)$ in each marginal. Then it holds

$$M_{\text{eff.}} = 1 + \sum_{j=2}^{m} \kappa_j, \quad \kappa_j = \frac{\log(\mathbb{P}_{\vartheta^*}(T_j \leq c_j | T_{j-1} \leq c_{j-1}))}{\log(1 - \alpha_{\text{loc.}})},$$

see Moskvina and Schmidt (2008).

(iii) Under the assumptions of part (ii), assume now that β_i applies for some $i > 2$. Then it holds

$$M_{\text{eff.}} = 1 + \xi(i) + \sum_{j=i}^{m} \kappa_j^{(i)},$$

where

$$\xi(i) = \sum_{\ell=2}^{i-1} \frac{\log(\gamma_{\ell,\ell}(c))}{\log(1 - \alpha_{\text{loc.}})}, \quad \kappa_j^{(i)} = \frac{\log(\gamma_{j,i}(c))}{\log(1 - \alpha_{\text{loc.}})},$$

see Dickhaus and Stange (2013).

4.4 Simultaneous Test Procedures in Terms of p-value Copulae

As shown by Dickhaus and Gierl (2013), simultaneous test procedures based on max-statistics can equivalently be described by p-value copulae, cf. also Sect. 2.2.4.

Theorem 4.8. *Let φ denote an STP for $(\mathcal{X}, \mathcal{F}, (\mathbb{P}_\vartheta)_{\vartheta \in \Theta}, \mathcal{H}_m)$ which is based on test statistics $(T_1, \ldots, T_m)$. Assume that the following three structural properties hold true.*

(S1) Any $\vartheta \in H_0$ is an LFC for the FWER of any STP for $(\mathcal{X}, \mathcal{F}, (\mathbb{P}_\vartheta)_{\vartheta \in \Theta}, \mathcal{H}_m)$ that is based on $(T_1, \ldots, T_m)$.

(S2) Every null hypothesis H_i, $1 \leq i \leq m$, is of the form $H_i : \{\theta_i(\vartheta) = \theta_i^*\}$, where $\theta : \Theta \to \Theta'$ denotes a derived parameter, i indexes components of θ, and the θ_i^* are fixed given values in Θ'.

(S3) The marginal cumulative distribution function of T_i under H_i, F_i (say), is continuous and strictly increasing.

Then, for arbitrary $\vartheta \in \Theta$ and $\vartheta^* \in H_0$, it holds

$$FWER_\vartheta(\varphi) \leq 1 - C_{\vartheta^*}(1 - \alpha_{loc.}^{(1)}, \ldots, 1 - \alpha_{loc.}^{(m)}), \tag{4.17}$$

with C_{ϑ^*} denoting the copula of the distributional transforms $(1 - p_i : 1 \leq i \leq m)$ of $T_1, \ldots, T_m$ under ϑ^*, $\alpha_{loc.}^{(i)} = 1 - F_i(c_\alpha)$, and c_α as in Definition 4.1.

To control the FWER at level α with the STP φ, one can therefore equivalently compare the (marginal) distributional transforms with a suitable $(1-\alpha)$-quantile of their copula under ϑ^*. As outlined in Sect. 2.2.5, this has the advantage that the local significance levels $(\alpha_{\mathrm{loc.}}^{(i)} : 1 \leq i \leq m)$ express the adjustment for multiplicity of the FWER level α explicitly. Furthermore, the rich and growing field of copula-based modeling of multivariate dependency structures becomes usable for multiple testing by means of Theorem 4.8. The possibility to employ copula-based models for constructing multiple tests has been mentioned in Sarkar (2008), but we are not aware of other references realizing this suggestion.

Example 4.2 (Dunnett contrasts under ANOVA1). Fix an integer k (number of treatment groups) and sample sizes $(n_i)_{1\leq i\leq k}$, and model the observation $x \in \mathcal{X} = \mathbb{R}^{\sum_{i=1}^{k} n_i}$ as a realization of $X = (X_{i,j} : 1 \leq i \leq k, 1 \leq j \leq n_i)$. In this, assume that

(i) all $X_{i,j}$ are stochastically independent,
(ii) $X_{i,j} \sim \mathcal{N}(\mu_i, 1)$ (or with unknown, but common variance).

The parameter of this model is the unknown mean vector $\mu = (\mu_1, \ldots, \mu_k)^\top \in \mathbb{R}^k$. Consider the "multiple comparisons with a control group" problem, i. e., the hypotheses $H_i : \mu_i = \mu_k$, $1 \leq i \leq k-1$, leading to $m = k-1$. Equivalently, we can express H_i as $\theta_i = 0$, where $\theta_i = \mu_i - \mu_k$ is a derived parameter. In a compact matrix notation, we can express $\mathcal{H}_{k-1} = (H_1, \ldots, H_{k-1})$ as

$$C_{\mathrm{Dunnett}}\, \mu = 0. \tag{4.18}$$

Line i of the latter system of equations is equal to H_i, $1 \leq i \leq k-1$. The contrast matrix C_{Dunnett} is Dunnett's contrast matrix with $k-1$ rows and k columns, where in each row j the jth entry equals $+1$, the kth entry equals -1 and all other entries are equal to zero. This is a classical multiple test problem which has been considered in the pioneering works of Charles W. Dunnett, cf. Dunnett (1955, 1964).

Denoting the empirical mean in group i by $\bar{X}_{i.}$, suitable (standard) test statistics for the two-sided comparisons defined by (4.18) are given by $|T_i|$, $1 \leq i \leq k-1$, where $T_i = \sqrt{n_i n_k/(n_i+n_k)}(\bar{X}_{i.} - \bar{X}_{k.})$. According to Theorem 4.4, the joint distribution of $T = (T_1, \ldots, T_{k-1})^\top$ is multivariate normal (or multivariate t) with a covariance matrix Σ which only depends on the sample sizes $n_1, \ldots, n_k$. More specifically, we have that $T \sim \mathcal{N}_{k-1}(\tilde{\mu}, \Sigma)$ with

$$\tilde{\mu}_i = \sqrt{\frac{n_i n_k}{n_i + n_k}}(\mu_i - \mu_k) \ \text{ and } \ \Sigma = D C_{\mathrm{Dunnett}} M C_{\mathrm{Dunnett}}^\top D,$$

where $D = \operatorname{diag}\left(\sqrt{\frac{n_i n_k}{n_i+n_k}} : 1 \leq i \leq k-1\right) \in \mathbb{R}^{k-1\times k-1}$ and $M = \operatorname{diag}(n_i^{-1} : 1 \leq i \leq k) \in \mathbb{R}^{k\times k}$. For ease of graphical illustration, let us now consider the case of $k = 3$ and, consequently, $m = 2$. We obtain that

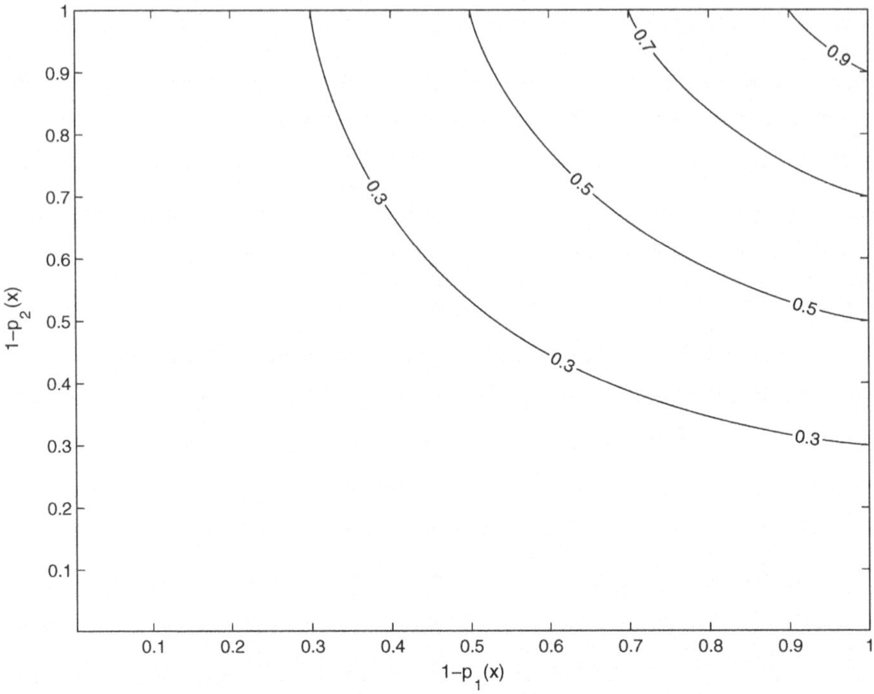

Fig. 4.2 Contour lines of C_{μ^*} in the case of $(n_1, n_2, n_3) = (90, 80, 70)$ for the STP from Example 4.2

$$\Sigma = \begin{pmatrix} 1 & \sqrt{\frac{n_1 n_2}{(n_1+n_3)(n_2+n_3)}} \\ \sqrt{\frac{n_1 n_2}{(n_1+n_3)(n_2+n_3)}} & 1 \end{pmatrix}.$$

Thus, since the joint distribution of T under the global hypothesis is exactly known here, the copula C_{μ^*} of the distributional transforms under H_0 can simply be calculated by transformation of measures. For $(u_1, u_2) \in [0, 1]^2$, we obtain

$$C_{\mu^*}(u_1, u_2) = F_{|T|}(\Phi^{-1}((u_1 + 1)/2), \Phi^{-1}((u_2 + 1)/2)),$$

where $F_{|T|}$ denotes the joint cdf. of the absolute values of T under μ^*, which is easily evaluable by numerical routines for multivariate normal distributions like the `mvtnorm` package in `R`, cf. Genz and Bretz (2009) and Sect. 2.2.5.

Figure 4.2 depicts contour lines of C_{μ^*} in the case of $(n_1, n_2, n_3) = (90, 80, 70)$ for contour levels 0.3, 0.5, 0.7, and 0.9. For control of the FWER at level $\alpha = 0.3$ (say) with the STP defined by T, Fig. 4.3 represents our findings from Theorem 4.8 graphically: An STP is constructed by determining the point of intersection of the diagonal on $[0, 1]^2$ with the contour line of C_{μ^*} at contour level $1 - \alpha$. Projection

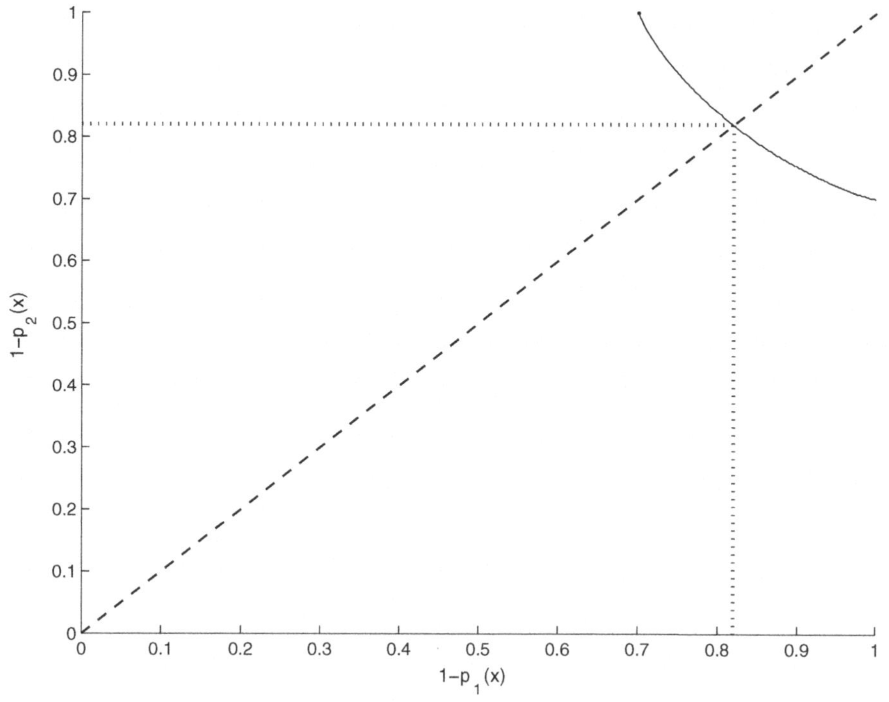

Fig. 4.3 Graphical representation of the construction of an STP according to Theorem 4.8

onto the coordinate axes yields the multiplicity-adjusted marginal copula arguments $1-\alpha^{(i)}_{\text{loc.}}$, $i = 1, 2$. In the example, one could consequently choose local significance levels $\alpha^{(1)}_{\text{loc.}} = \alpha^{(2)}_{\text{loc.}} = 0.1812$.

Since every bivariate $(1-\alpha)$-quantile of C_{ϑ^*} is a solution to the problem of STP construction according to Theorem 4.8, Fig. 4.3 furthermore shows how an importance weighting of the individual hypotheses can be incorporated straightforwardly into the method: the only thing that has to be changed is the slope of the line through the origin.

4.5 Exploiting the Topological Structure of the Sample Space via Random Field Theory

In many applications from modern life sciences, the observational units are geometrically related, and the scientific questions relate to aggregated quantities defining topological regions. One example is functional magnetic resonance imaging (fMRI), where the blood oxygen level is measured in voxels, but brain activity is assumed to be constituted by spatial clusters of voxels (brain regions). We will provide more details in Chap. 11. Another example is given by association or quantitative trait loci analyses in genetics, where genetic loci are used as markers for genomic regions, cf. Chaps. 9 and 10. A mathematical tool for incorporating the topological structure of the sample space into the construction of simultaneous test procedures for such types of problems is the theory of random fields.

For the purpose of indicating the incorporation of the geometry of the sample space into the analysis, we denote in this section the index set for the system of hypotheses $\mathscr{H}$ which is of interest by $\mathscr{S}$ (a finite-dimensional manifold). More specifically, we assume that the data sample results in a realization of a stochastic process (called random field) T with values in $\mathbb{R}^{\mathscr{S}}$, meaning that a real number (the value of a test statistic) can be observed at every location $s \in \mathscr{S}$, when hypotheses are formulated for every such location. An important branch of random field theory is considered with computing or approximating excursion probabilities of the form

$$\mathbb{P}\left(\max_{s \in \mathscr{S}} T(s) \geq t\right), \quad t \in \mathbb{R}, \tag{4.19}$$

where $\mathbb{P}$ is the probability measure driving the random field T. In the context of simultaneous test procedures, $\mathbb{P}$ will be a probability measure corresponding to the global hypothesis $H_0 = \bigcap_{s \in \mathscr{S}} H_s$. For instance, in the fMRI example, $s \in \mathscr{S}$ would indicate a spatial position in the brain, while it would indicate a locus on the genome in the aforementioned examples from genetics. The purpose then is to calibrate $t = t(\alpha)$ with respect to (multiple) type I error control or, equivalently, to evaluate the probability in (4.19) at the observed data points in order to provide a multiplicity-adjusted p-value for each hypothesis H_s, $s \in \mathscr{S}$. The theory is well-developed for cases in which each $T(s)$ is Gaussian or its distribution is related to the normal distribution (for example, Student's t, chi-squared, or Fisher's F) under H_s. One particularly useful approximation of the excursion probability in (4.19) is given by the so-called "Euler characteristic heuristic" (cf., e. g., Sect. 5.1 in Adler and Taylor (2011)), meaning that

$$\mathbb{P}\left(\max_{s \in \mathscr{S}} T(s) \geq t\right) \approx \mathbb{E}[\chi\{s \in \mathscr{S} : T(s) \geq t\}]. \tag{4.20}$$

The quantity $\chi\{s \in \mathscr{S} : T(s) \geq t\}$ is the (random) Euler characteristic (EC) of the excursion set of T over the threshold t on $\mathscr{S}$. Although the EC itself is a complicated object depending on the geometry of $\mathscr{S}$ and the distributional properties of T, its expectation can in Gaussian or Gaussian-related cases be computed explicitly by the Gaussian kinematic formula derived by Taylor (2006). In particular, the following result holds true.

Lemma 4.3. (Theorem 4.8.1 in Adler and Taylor (2011)). *Assume that $\mathscr{S}$ and $D \subset \mathbb{R}$ are regular stratified manifolds and that $T : \mathscr{S} \to \mathbb{R}$ is a Gaussian random field on $\mathscr{S}$ with mean zero and constant unit variance. Then it holds*

$$\mathbb{E}[\chi\{\mathcal{S} \cap T^{-1}(D)\}] = \sum_{j=0}^{\dim \mathcal{S}} (2\pi)^{-j/2} \mathcal{L}_j(\mathcal{S}) \mathcal{M}_j^{\mathcal{N}(0,1)}(D), \tag{4.21}$$

where $\mathcal{L}_j(\mathcal{S})$ denotes the j-th Lipschitz-Killing curvature of $\mathcal{S}$ with respect to the variogram metric induced by T and $\mathcal{M}_j^{\mathcal{N}(0,1)}$ the j-th Gaussian Minkowski functional on $\mathbb{R}$.

For a general probability measure $\mathbb{P}$ on $\mathbb{R}$, the numbers $\mathcal{M}_j^{\mathbb{P}}(D)$ can implicitly be defined as the coefficients of the power series expansion of the $\mathbb{P}$-volume of the δ-tube around D, i. e., via the equation

$$\mathbb{P}(\text{Tube}(D, \delta)) = \sum_{j=0}^{\infty} \frac{\delta^j}{j!} \mathcal{M}_j^{\mathbb{P}}(D), \text{ where } \text{Tube}(D, \delta) = \{u \in \mathbb{R} : \inf_{v \in D} |u - v| \leq \delta\}.$$

For the special case of excursion sets as in (4.19), we notice that

$$\mathcal{M}_j^{\mathcal{N}(0,1)}([t, \infty)) = \frac{1}{\sqrt{2\pi}} H_{j-1}(t) \exp(-t^2/2),$$

with H_ℓ, $\ell \geq 0$, denoting the ℓ-th Hermite polynomial and H_{-1} defined by

$$H_{-1}(t) = \sqrt{2\pi} \exp(t^2/2)(1 - \Phi(t)),$$

leading to $\mathcal{M}_0^{\mathcal{N}(0,1)}([t, \infty)) = 1 - \Phi(t)$ and

$$\mathbb{E}[\chi\{s \in \mathcal{S} : T(s) \geq t\}] = \exp(-t^2/2) \sum_{j=0}^{\dim \mathcal{S}} (2\pi)^{-\frac{j+1}{2}} \mathcal{L}_j(\mathcal{S}) H_{j-1}(t). \tag{4.22}$$

Furthermore, as explained in Sect. 5.2 of Adler and Taylor (2011), formula (4.21) remains to hold true if the distribution of $T(s)$ is not Gaussian, but Gaussian-related, with the only difference that the set D has to be replaced by a different set D' which expresses the form of relatedness of the distribution of $T(s)$ to the standard normal distribution. In essence, one has to compute the $\mathcal{N}(0, 1)$-volume of a geometrically different object than just a half-open interval, which can in practically relevant cases be done by means of standard methods from stochastics; see, for instance, Worsley (1994). Hence, in practice all that remains is to calculate or to estimate the Lipschitz-Killing curvatures $\mathcal{L}_j(\mathcal{S})$, $0 \leq j \leq \dim \mathcal{S}$, appearing in (4.21) and (4.22). These numbers depend on the geometry of the manifold $\mathcal{S}$ and on the local correlation structure among $(T(s) : s \in \mathcal{S})$. We will return to the estimation task for the $\mathcal{L}_j(\mathcal{S})$ in the special context of fMRI analyses in Chap. 11.

Acknowledgments Parts of this chapter originated from joint work with Jens Stange. Figure 4.1 has been adapted from a presentation slide of Edgar Brunner. Section 4.4 has been the topic of Jakob Gierl's diploma thesis from which I adapted Figs. 4.2 and 4.3. Thomas Royen taught me everything I know about multivariate chi-square distributions. I am grateful to Mette Langaas and Øyvind Bakke for inviting me and for their hospitality during my visit to Norwegian University of Science and Technology (NTNU), for many fruitful discussions and for critical reading.

References

Adler RJ, Taylor JE (2011) Topological complexity of smooth random functions. École d'Été de Probabilités de Saint-Flour XXXIX-2009. Lecture Notes in Mathematics 2019. Springer, Berlin. doi:10.1007/978-3-642-19580-8

Block HW, Costigan T, Sampson AR (1992) Product-type probability bounds of higher order. Probab Eng Inf Sci 6(3):349–370. doi:10.1017/S0269964800002588

Conneely KN, Boehnke M (2007) So many correlated tests, so little time! Rapid adjustment of P values for multiple correlated tests. Am J Hum Genet 81(6):1158–1168

Das Gupta S, Olkin I, Savage L, Eaton M, Perlman M, Sobel M (1972) Inequalities on the probability content of convex regions for elliptically contoured distributions. In: Proceedings of 6th Berkeley symposium on mathematical statistics and probability, University of California 1970, vol 2, pp 241–265

Dickhaus T, Gierl J (2013) Simultaneous test procedures in terms of p-value copulae. Global Science and Technology Forum (GSTF). In: Proceedings on the 2nd annual international conference on computational mathematics, computational geometry & statistics (CMCGS 2013), vol 2. pp 75–80

Dickhaus T, Stange J (2013) Multiple point hypothesis test problems and effective numbers of tests for control of the family-wise error rate. Calcutta Statist Assoc Bull, to appear

Dunnett CW (1955) A multiple comparison procedure for comparing several treatments with a control. J Am Stat Assoc 50:1096–112. doi:10.2307/2281208

Dunnett CW (1964) New tables for multiple comparisons with a control. Biometrics 20:482–491

Efron B (1997) The length heuristic for simultaneous hypothesis tests. Biometrika 84(1):143–157. doi:10.1093/biomet/84.1.143

Fahrmeir L, Hamerle A (1984) Multivariate statistische Verfahren. Unter Mitarbeit von Walter Häußler, Heinz Kaufmann, Peter Kemény, Christian Kredler, Friedemann Ost, Heinz Pape, Gerhard Tutz. Berlin-New York: Walter de Gruyter.

Gabriel KR (1969) Simultaneous test procedures - some theory of multiple comparisons. Ann Math Stat 40:224–250. doi:10.1214/aoms/1177697819

Galambos J, Rényi A (1968) On quadratic inequalities in probability theory. Stud Sci Math Hung 3:351–358

Genz A, Bretz F (2009) Computation of multivariate normal and *t* probabilities. Lecture Notes in Statistics, vol 195. Springer, Berlin. doi:10.1007/978-3-642-01689-9

Glaz J, Johnson BM (1984) Probability inequalities for multivariate distributions with dependence structures. J Am Stat Assoc 79:436–440. doi:10.2307/2288287

Hothorn T, Bretz F, Westfall P (2008) Simultaneous inference in general parametric models. Biom J 50(3):346–363

Hunter D (1976) An upper bound for the probability of a union. J Appl Probab 13:597–603. doi:10.2307/3212481

Jensen DR (1970) A generalization of the multivariate Rayleigh distribution. Sankhyā Ser A 32(2):193–208

Karlin S, Rinott Y (1980) Classes of orderings of measures and related correlation inequalities. I. multivariate totally positive distributions. J Multivariate Anal 10:467–498

Kotz S, Nadarajah S (2004) Multivariate *t* distributions and their applications. Cambridge University Press, Cambridge

Moskvina V, Schmidt KM (2008) On multiple-testing correction in genome-wide association studies. Genet Epidemiol 32:567–573

Naiman D, Wynn H (2005) The algebra of Bonferroni bounds: discrete tubes and extensions. Metrika 62(2–3):139–147. doi:10.1007/s00184-005-0403-2

Naiman DQ, Wynn HP (1992) Inclusion-exclusion-Bonferroni identities and inequalities for discrete tube-like problems via Euler characteristics. Ann Stat 20(1):43–76. doi:10.1214/aos/1176348512

Naiman DQ, Wynn HP (1997) Abstract tubes, improved inclusion-exclusion identities and inequalities and importance sampling. Ann Stat 25(5):1954–1983. doi:10.1214/aos/1069362380

Naiman DQ, Wynn HP (2001) Improved inclusion-exclusion inequalities for simplex and orthant arrangements. J Inequal Pure Appl Math 2(2):Paper No.18

Ninomiya Y, Fujisawa H (2007) A conservative test for multiple comparison based on highly correlated test statistics. Biometrics 63(4):1135–1142. doi:10.1111/j.1541-0420.2007.00821.x

Rényi A (1961) Eine allgemeine Methode zum Beweis von Gleichungen der Wahrscheinlichkeitsrechnung mit einigen Anwendungen. Magyar Tud Akad Mat Fiz Tud Oszt Közl 11:79–105

Royen T (2007) Integral representations and approximations for multivariate gamma distributions. Ann Inst Stat Math 59(3):499–513. doi:10.1007/s10463-006-0057-5

Sarkar SK (2008) Rejoinder: On methods controlling the false discovery rate. Sankhyā 70(2, Ser. A):183–185

Šidák Z (1967) Rectangular confidence regions for the means of multivariate normal distributions. J Am Stat Assoc 62:626–633. doi:10.2307/2283989

Taylor JE (2006) A Gaussian kinematic formula. Ann Probab 34(1):122–158. doi:10.1214/009117905000000594

Timm NH (2002) Applied multivariate analysis. Springer, New York

Tong Y (1990) The multivariate normal distribution. Springer series in statistics. Springer, New York

Westfall PH, Young SS (1993) Resampling-based multiple testing: examples and methods for p-value adjustment. Wiley series in probability and mathematical statistics. Applied probability and statistics. Wiley, New York

Worsley K (1982) An improved Bonferroni inequality and applications. Biometrika 69:297–302. doi:10.1093/biomet/69.2.297

Worsley K (1994) Local maxima and the expected Euler characteristic of excursion sets of χ^2, F and t fields. Adv Appl Probab 26(1):13–42. doi:10.2307/1427576

Chapter 5
Stepwise Rejective Multiple Tests

Abstract We are considered with step-up-down tests for control of the family-wise error rate (FWER) and the false discovery rate (FDR), respectively. FWER-controlling step-down tests are derived by applying the closed test principle in connection with Bonferroni and Šidák corrections. FWER-controlling step-up tests are based on Simes' global test, again in connection with the closed test principle. The theory of step-up tests for control of the FDR is developed by taking the famous linear step-up procedure by Benjamini and Hochberg (1995) as the starting point. Explicitly and implicitly adaptive FDR-controlling step-up-down tests are finally treated. A precise discussion about the underlying (dependency) assumptions of each procedure is provided and summarized in a table, guiding the user to appropriate methods for an actual multiple test problem at hand.

In this chapter, we investigate margin-based step-up-down tests, cf. Definition 3.1. They are particularly suitable for large systems of hypotheses when modeling or reliably estimating the full joint distribution of test statistics or p-values, respectively, is infeasible due to the "curse of dimensionality", meaning that the parameter space is of higher dimensionality than the sample size. If only marginal distributions of test statistics are modeled explicitly, one is particularly interested in "generic" multiple test procedures that provide the desired error control over a broad class of joint distributions. However, sometimes some qualitative assumption regarding the dependency structure is available and the multiple tests are constructed to be generic over the subclass of all multivariate distributions fulfilling this assumption, which typically leads to more powerful multiple tests in comparison to cases with completely unspecified dependencies.

For orientation, the following table lists all multiple test procedures treated in this chapter. They can be systematized by their structure (step-up (SU), step-down (SD), step-up-down (SUD)), by the type of error control they provide, and by their respective assumptions regarding dependency among p-values.

T. Dickhaus, *Simultaneous Statistical Inference*,
DOI: 10.1007/978-3-642-45182-9_5,

Table 5.1 Overview of stepwise rejective multiple tests

Multiple test	Structure	Error control	Dependency
Bonferroni-Holm	SD	FWER	Arbitrary
Šidák-Holm	SD	FWER	Independence
Hommel	SU	FWER	MTP_2
Hochberg	SU	FWER	MTP_2
Rom	SU	FWER	Independence
Benjamini-Hochberg	SU	FDR	PRDS
Storey-Taylor-Siegmund	SU	FDR	weak dependency
Benjamini-Yekutieli	SU	FDR	Arbitrary
Blanchard-Roquain	SU	FDR	Arbitrary
AORC	SUD	FDR	Independence

We will define the concepts of dependency listed in Table 5.1 in Sect. 5.1 below. From the practical perspective, it is essential to notice that type I error control of the respective procedures can only be guaranteed if the corresponding assumptions regarding dependency of test statistics or p-values, respectively, hold true. Hence, Table 5.1 may be used as a guideline to choose a multiple test procedure for a particular problem at hand. We will discuss the appropriateness of the positive dependency assumptions given in Table 5.1 in the context of genetic association studies (Sect. 9.5), in the context of gene expression analysis (Sect. 10.2) and in the context of functional magnetic resonance imaging (Sect. 11.1); see also Sect. 12.2 for an application in proteomics.

5.1 Some Concepts of Dependency

Definition 5.1 (MTP_2 and PRDS). Let $(\mathscr{X}, \mathscr{F}, \mathbb{P})$ be a probability space and let $T = (T_1, \ldots, T_m)^\top : \mathscr{X} \to S^m$ be a random vector, where $(S, \mathscr{S})$ is a measurable space with S a subset of $\mathbb{R}$.

(i) The vector T is called multivariate totally positive of order 2 (MTP_2), if its distribution $\mathbb{P}^T$ on $(S^m, \mathscr{S}^{\otimes m})$ has a probability density function $f : S^m \to [0, \infty)$ with respect to a measure $\sigma^{\otimes m}$, such that for all $u, v \in S^m$:

$$f(u) \cdot f(v) \leq f(\min(u, v)) \cdot f(\max(u, v)),$$

where the minimum or maximum, respectively, is being taken component-wise.

(ii) The vector T is called positive regression dependent on a subset I_0 of the set of indices $\{1, \ldots, m\}$ (PRDS on I_0), if for every increasing set $D \subset S^m$ and for every index $i \in I_0$

$$\mathbb{P}(T \in D \mid T_i = u) \text{ is non-decreasing in } u.$$

Therein, the set D is called increasing if $u_1 \in D$ and $u_2 \geq u_1$ (jointly) imply $u_2 \in D$.

In Definition 5.1, the dependency concepts are formulated in terms of test statistics $T_1, \ldots, T_m$. However, we will use them in this chapter mainly in connection with p-values $p_1, \ldots, p_m$. Since p-values are typically deterministic, monotone transformations of test statistics, this distinction is essentially void, meaning that the dependency concept holds for the p-values if and only if it holds for the test statistics.

Examples of MTP$_2$ and PRDS distributions are given by, e.g., Karlin and Rinott (1980); Sarkar and Chang (1997); Sarkar (1998); Benjamini and Yekutieli (2001), and Sarkar (2002). For the multivariate t-distribution, see also the recent result by Block et al. (2013). Here, we mention three examples of MTP$_2$ families explicitly.

Lemma 5.1.

(a) Let $X = (X_1, \ldots, X_m)$ denote a multivariate Gaussian random vector, i. e., $X \sim \mathcal{N}_m(\mu, \Sigma)$ with $\mu \in \mathbb{R}^m$ and $\Sigma \in \mathbb{R}^{m \times m}$. If Σ is positive definite and all its entries are non-negative, then X is *MTP*$_2$.
(b) Let $X = (X_1, \ldots, X_m)$ denote a centered multivariate Gaussian random vector, $X \sim \mathcal{N}_m(0, \Sigma)$ with Σ positive definite and let $|X| = (|X_1|, \ldots, |X_m|)$. Then, $|X|$ is *MTP*$_2$ if and only if there exists a diagonal matrix D with diagonal elements ± 1 such that the off-diagonal elements of $-D\Sigma^{-1}D$ are all non-negative.
(c) Let $Q = (Q_1, \ldots, Q_m)$ follow a multivariate central chi-squared distribution with ν degrees of freedom in every marginal. Under exchangeability (entailing equi-correlation) in the sense that each Q_j can be represented as $Q_j = X_j + X_0$ for stochastically independent, chi-squared distributed variates $X_0, X_1, \ldots, X_m$, Q is *MTP*$_2$.

Proof. Part (a) follows from Sect. 4.3.3. in Tong (1990), part (b) is Theorem 3.1 in Karlin and Rinott (1980) and part (c) is a consequence of Example 3.5. in Karlin and Rinott (1980). □

Lemma 5.2 (cf., e.g., Hu et al. (2006)). *MTP*$_2$ *implies PRDS on any subset of* $\{1, \ldots, m\}$.

Definition 5.2 (Weak dependency, cf. Storey et al. (2004)). Let $p_1, \ldots, p_m$ denote (random) marginal p-values for a multiple test problem $(\mathcal{X}, \mathcal{F}, (\mathbb{P}_\vartheta)_{\vartheta \in \Theta}, \mathcal{H}_m)$ with index set I_0 of true hypotheses and index set $I_1 = \{1, \ldots, m\} \setminus I_0$ of false hypotheses in $\mathcal{H}_m$ under ϑ, and $m_0 = |I_0|, m_1 = |I_1|$. Then, $p_1, \ldots, p_m$ are called weakly dependent under ϑ, if $\pi_0 = \lim_{m \to \infty} m_0/m$ exists and

$$m_0^{-1} \sum_{i \in I_0} \mathbf{1}_{[0,t]}(p_i) \to F_0(t), \quad m \to \infty, \tag{5.1}$$

$$m_1^{-1} \sum_{i \in I_1} \mathbf{1}_{[0,t]}(p_i) \to F_1(t), \quad m \to \infty, \tag{5.2}$$

where convergence in (5.1) and (5.2) is uniformly in $t \in [0, 1]$ and almost surely, and F_0 and F_1 are continuous functions with $0 < F_0(t) \leq t$ for all $t \in (0, 1]$.

5.2 FWER-Controlling Step-Down Tests

FWER-controlling step-down tests naturally arise in connection with the closed test principle by a technique called "shortcut". Plainly phrased, a shortcut of a closed test procedures avoids explicit testing of all intersection hypotheses in the $\cap$-closed system of hypotheses $\bar{\mathscr{H}}$ induced by $\mathscr{H}$. This is done by traversing the closure $\bar{\mathscr{H}}$ in a group-wise manner. A nice general theory of shortcuts can be found in Hommel et al. (2007). Here, we demonstrate the principle by identifying the Bonferroni-Holm test and the Šidák-Holm test as shortcuts of the Bonferroni and Šidák single-step test, respectively (see Sect. 3.1.1).

Definition 5.3 (Bonferroni-Holm and Šidák-Holm tests, see Holm (1977, 1979)). Let $(\mathscr{X}, \mathscr{F}, \mathscr{P}, \mathscr{H}_m)$ denote a multiple test problem with finite system of hypotheses $\mathscr{H}_m = \{H_i, i \in I = \{1, \ldots, m\}\}$. Without loss of generality, assume that $H_1, \ldots, H_k$ are the elementary hypotheses in $\mathscr{H}_m$, for some $1 \leq k \leq m$. Assume that p-values $p_i : 1 \leq i \leq k$ are available for every marginal test problem H_i versus K_i, $i \in \{1, \ldots, k\} = I_k$ (say). Let $p_{1:k} \leq p_{2:k} \leq \cdots \leq p_{k:k}$ denote the order statistics of these p-values and $H_{1:k}, \ldots, H_{k:k}$ the correspondingly sorted elementary hypotheses in $\mathscr{H}_m$. Define, for $i \in I_k$,

$$\alpha_i = \begin{cases} 1 - (1 - \alpha)^{1/i}, & \text{if } (p_i(X), i \in I_k) \text{ are stochastically independent,} \\ \alpha/i, & \text{otherwise.} \end{cases}$$

Then, the Šidák-Holm test (independent case) or the Bonferroni-Holm test (case of general dependencies) φ^{Holm} (say) rejects (exactly) the elementary hypotheses $H_{1:k}, \ldots, H_{i^*:k}$, where

$$i^* = \max\{i \in I_k : p_{j:k} \leq \alpha_{k-j+1} \text{ for all } j = 1, \ldots, i\}.$$

Furthermore, an intersection hypothesis H_ℓ, where $\ell \in \{k+1, \ldots, m\}$, gets rejected if and only if at least one elementary hypothesis which is used for intersection has been rejected.

Theorem 5.1. *The multiple test φ^{Holm} for $(\mathscr{X}, \mathscr{F}, \mathscr{P}, \mathscr{H}_m)$ is coherent and consonant, and it strongly controls the FWER at level α. If the system of hypotheses is complete, meaning that $|\bar{\mathscr{H}}_m| = 2^k - 1$, then φ^{Holm} is equivalent to a closed test $\bar{\varphi}$ (say). The multiple test φ which induces $\bar{\varphi}$ tests every intersection hypothesis H_J in $\bar{\mathscr{H}}_m$ by applying a "local" Bonferroni correction or Šidák correction, respectively, to the p-values p_i, $i \in J$, where J denotes the set of indices of the elementary hypotheses used for intersection.*

Proof. Coherence and consonance of φ^{Holm} follow directly from Definition 5.3. For proving strong FWER control of φ^{Holm}, we investigate the closed test $\bar{\varphi}$ and show that it rejects all hypotheses which are rejected by φ^{Holm}, possibly more. Since $\bar{\varphi}$ is strongly FWER-controlling, this implies strong FWER control of φ^{Holm}. For ease of argumentation, we write $\bar{\mathscr{H}}_m$ in the form $\bar{\mathscr{H}}_m = \{H_J : J \in \bar{I}\}$ for the appropriate index set $\bar{I} \subseteq 2^{\{1,\ldots,k\}}$, where $H_J = \bigcap_{j \in J} H_j$. Obviously, for an elementary hypothesis H_r, $H_r \supseteq H_J$ implies $r \in J$. The multiple test φ which induces $\bar{\varphi}$ is with this notation given by ("local" Bonferroni or Šidák correction)

$$\varphi_J(x) = \begin{cases} 1, & \text{if } \min_{j \in J} p_j \leq \alpha_{|J|}, \\ 0, & \text{if } \min_{j \in J} p_j > \alpha_{|J|}. \end{cases}$$

From Lemma 1.1 it follows that φ is a multiple test at local level α. Following the closed test principle from Theorem 3.4, we get that $\bar{\varphi}$ is defined by

$$\bar{\varphi}_J(x) = \begin{cases} 1, & \text{if} \, \forall L \supseteq J, L \in \bar{I} : \varphi_L(x) = 1, \\ 0, & \text{otherwise.} \end{cases}$$

Without loss of generality, assume $p_1 \leq \cdots \leq p_k$. Abbreviating $\bar{\varphi}_i := \bar{\varphi}_{\{i\}}$, $i \in I_k$, it holds that $\bar{\varphi}_i(x) = 1$ if and only if for all $J \in \bar{I}$ with $i \in J$, $\min_{j \in J} p_j \leq \alpha_{|J|}$. Equivalently, it holds

$$\forall r \in \{1, \ldots, k\} : \forall J \in \bar{I} \text{ with } i \in J \text{ and } |J| = r : \min_{j \in J} p_j \leq \alpha_r. \tag{5.3}$$

However, (5.3) is also fulfilled for φ^{Holm}. To see this, assume that $p_i \leq \alpha_{k-i+1}$ for all $1 \leq i \leq i^*$. Then, it obviously holds for all indices $1 \leq i \leq i^*$ of rejected elementary hypotheses in $\mathscr{H}_m$ that

$$\forall r \in \{1, \ldots, k\} : \forall J \in \bar{I} \,\text{with}\, i \in J \,\text{and}\, |J| = r : \min_{j \in J} p_j = p_{\min\{j \in J\}} \leq \alpha_{k-\min\{j \in J\}+1}.$$

Furthermore, $\alpha_{k-\min\{j \in J\}+1} \leq \alpha_r$, because $\min\{j \in J\} \leq k - r + 1$ and α_ℓ decreases in ℓ. Finally, easy combinatorial considerations show that $\min\{j \in J\} = k - r + 1$ if $\bar{\mathscr{H}}_m$ is complete, completing the proof. □

Remark 5.1.

(a) The condition $|\bar{\mathscr{H}}_m| = 2^k - 1$ is also occasionally referred to as the "free combinations" condition regarding the hypotheses in $\mathscr{H}_m$, see, e.g., Troendle and Westfall (2011).

(b) Holm's tests are uniform improvements over the corresponding Bonferroni or Šidák correction with respect to multiple power. However, constructing compatible simultaneous confidence regions is much less straightforward for a step-down test like φ^{Holm} than for a single-step (for instance, Bonferroni or Šidák) test. The latter task has been independently addressed only recently by Strassburger and Bretz (2008) and by Guilbaud (2008), see also Guilbaud (2012) and Guilbaud and Karlsson (2011).

5.3 FWER-Controlling Step-Up Tests

The starting point for the development of FWER-controlling step-up tests is Simes' global test.

Lemma 5.3 (Simes (1986)). *Let $U_1, \ldots, U_m$ denote stochastically independent, identically UNI[0, 1]-distributed random variables and $U_{1:m} \leq \cdots \leq U_{m:m}$ their order statistics. Define $\alpha_{i:m} = i\alpha/m$, $1 \leq i \leq m$, for $\alpha \in [0, 1]$. Then it holds*

$$\mathbb{P}(U_{1:m} > \alpha_{1:m}, \ldots, U_{m:m} > \alpha_{m:m}) = 1 - \alpha. \tag{5.4}$$

The constants $\alpha_{i:m} = i\alpha/m$, $1 \leq i \leq m$, are referred to as Simes' critical values and play an important role in multiple testing, see also Definition 5.6. Recall from Chap. 2 that valid p-values are stochastically lower-bounded by UNI [0, 1] under null hypotheses. If the distribution of the U_i is stochastically larger than *UNI*[0, 1], then $1 - \alpha$ is a lower bound for the probability on the left-hand side of (5.4).

Simes' result has been extended by Sarkar (1998) to treat MTP_2 families of p-values.

Lemma 5.4 (Sarkar (1998)). *Under the assumptions of Lemma 5.3, but with joint independence of $U_1, \ldots, U_m$ replaced by requiring that $U_1, \ldots, U_m$ are MTP_2, it holds*

$$\mathbb{P}(U_{1:m} > \alpha_{1:m}, \ldots, U_{m:m} > \alpha_{m:m}) \geq 1 - \alpha. \tag{5.5}$$

Corollary 5.1 (Simes' global test). *Consider a multiple test problem $(\mathscr{X}, \mathscr{F}, \mathscr{P}, \mathscr{H}_m)$ and marginal p-values p_i, $i \in I$, which are MTP_2 under the global hypothesis $H_0 = \bigcap_{i=1}^m H_i$. Then, a level α test for H_0 is given by*

$$\varphi^{Simes}(x) = 1 \quad \Leftrightarrow \quad \exists i \in I : p_{i:m} \leq \frac{i}{m}\alpha.$$

We refer to φ^{Simes} as Simes' global test.

Remark 5.2. Under the assumptions of Corollary 5.1, Simes' global test can also be used as the basis for a closed test procedure $\bar{\varphi} = (\bar{\varphi}_i, i \in I)$ for $\mathscr{H}_m$. To this end, denote an intersection hypothesis by $H_J = \bigcap_{j \in J} H_j$ and let $p_{1:J} \leq \cdots \leq p_{|J|:J}$ denote the order statistics of $(p_j : j \in J)$. Then, the closed test procedure $\bar{\varphi}$ is given by

$$\bar{\varphi}_i(x) = 1 \quad \Leftrightarrow \quad \forall J \ni i : \exists j \in J : p_{j:J} \leq \alpha_{j:|J|}.$$

A shortcut version of the closed test procedure outlined in Remark 5.2 has been derived by Hommel (1988).

Definition 5.4 (Hommel (1988)). Under the assumptions of Corollary 5.1, assume without loss of generality that $p_1 \leq \cdots \leq p_m$. Let

$$J^* = \{i \in I : p_{m-i+k} > \frac{k\alpha}{i} \text{for all } k = 1, \ldots, i\},$$

$$j^* = \begin{cases} \max\{i : i \in J^*\}, & \text{if } J^* \neq \emptyset, \\ 1, & \text{if } J^* = \emptyset. \end{cases}$$

Then, the test $\varphi^{Hommel} = (\varphi_i^{Hommel}, i \in I)$ is given by

$$\varphi_i^{Hommel}(x) = \begin{cases} 1, & i \leq m^* := \max\{j : p_j \leq \alpha/j^*\}, \\ 0, & \text{otherwise.} \end{cases}$$

As shown by Hommel (1988), it holds that $\varphi_i^{Hommel} = \bar{\varphi}_i$ for all $i \in I$, where $\bar{\varphi}_i$ is as in Remark 5.2. However, the formulation via φ^{Hommel} reduces the computational effort in comparison to the explicit consideration of all intersection hypotheses. Notice that, although derived from the closed test principle, φ^{Hommel} has the structure of a step-up test.

A second example of an FWER-controlling step-up test is the procedure by Hochberg (1988).

Definition 5.5 (Hochberg (1988)). Let

$$\tilde{m} = \max\{i \in I : p_{i:m} \leq \frac{\alpha}{m - i + 1}\}.$$

Then the step-up test of Hochberg (1988), say $\varphi^{Hochberg} = (\varphi_i^{Hochberg}, i \in I)$, is given by

$$\varphi_i^{Hochberg}(x) = 1 \quad \Leftrightarrow \quad p_i \leq p_{\tilde{m}:m}.$$

Remark 5.3.

(a) Regarded as a function of the data, $\varphi^{Hochberg}$ is component-wise not larger than φ^{Hommel}. Hence, $\varphi^{Hochberg}$ controls the FWER at level α under the MTP_2 assumption regarding the joint distribution of $p_1, \ldots, p_m$.

(b) The multiple test $\varphi^{Hochberg}$ employs the same set of critical values as the Bonferroni-Holm test which is FWER-controlling under arbitrary dependency among the p-values. However, the MTP_2 assumption allows to change the structure of the test from step-down (Bonferroni-Holm) to the more powerful step-up (Hochberg), cf. Lemma 3.1.

As the final example, we derive a step-up test with FWER exactly equal to α for independent p-values which are uniformly distributed on [0, 1] under null hypotheses.

To this end, we recall the following result concerning the recursive computation of the joint cdf of order statistics of iid. random variables.

Lemma 5.5 (see, e.g., Shorack and Wellner (1986), p. 366). *Let $X_1, \ldots, X_m$ denote iid. random variables with cdf F of X_1, such that $F(x) = \mathbb{P}(X_1 \leq x), x \in \mathbb{R}$, and denote their order statistics by $X_{1:m}, \ldots, X_{m:m}$. Let $c_1 \leq \cdots \leq c_m$ denote real constants and let $\alpha_j = 1 - F(c_j)$, $j = 1, \ldots, m$. Let $F_j(c_1, \ldots, c_j) = \mathbb{P}(X_{1:j} \leq c_1, \ldots, X_{j:j} \leq c_j)$, $j = 1, \ldots, m$, with $F_0 \equiv 1$. Then it holds*

$$F_m(c_1, \ldots, c_m) = 1 - \sum_{j=0}^{m-1} \binom{m}{j} F_j(c_1, \ldots, c_j) \alpha_{j+1}^{m-j}. \tag{5.6}$$

Proof.

$$\begin{aligned} F_m(c_1, \ldots, c_m) &= \mathbb{P}(X_{1:m} \leq c_1, \ldots, X_{m:m} \leq c_m) = 1 - \mathbb{P}(\exists j \in \{1, \ldots, m\} : X_{j:m} > c_j) \\ &= 1 - \sum_{j=0}^{m-1} \mathbb{P}(X_{1:j} \leq c_1, \ldots, X_{j:j} \leq c_j, X_{j+1:m} > c_{j+1}) \\ &= 1 - \sum_{j=0}^{m-1} F_j(c_1, \ldots, c_j) \alpha_{j+1}^{m-j} \binom{m}{j}. \end{aligned}$$ □

We will use this result, with $X_i = 1 - p_i$, $1 \leq i \leq m$, in order to compute exact critical values for an FWER-controlling step-up test under joint independence of p-values recursively. Notice that the ordering of the p-values $p_1, \ldots, p_m$ is reverse to the ordering of the so defined variates $X_1, \ldots, X_m$. The following theorem guarantees the existence of a solution for any value of m.

Theorem 5.2 (Dalal and Mallows (1992)). *Under the assumptions of Lemma 5.5, assume that F is a continuous cdf on $\mathbb{R}$ and that $(X_m)_{m \in \mathbb{N}}$ is an iid sequence with $X_1 \sim F$. Then, for all $\alpha \in (0, 1)$, there exists a sequence $(c_m)_{m \in \mathbb{N}}$ of real numbers fulfilling*

$$\forall i \in \mathbb{N} : c_i < c_{i+1} \text{ and } \forall m \in \mathbb{N} : F_m(c_1, \ldots, c_m) = 1 - \alpha, \tag{5.7}$$

where F_m is as in Lemma 5.5.

For the computation of $\alpha_j, j \in I$, notice that, trivially, $\alpha_1 = \alpha$. Requiring

$$F_m(c_1, \ldots, c_m) = 1 - \alpha \text{ and} \tag{5.8}$$

$$F_j(c_1, \ldots, c_j) = 1 - \alpha \text{ for all } j = 1, \ldots, m - 1, \tag{5.9}$$

we conclude with (5.6) that, for all $m \geq 2$,

$$1-\alpha = 1-\sum_{j=0}^{m-1}\binom{m}{j}F_j(c_1,\ldots,c_j)\alpha_{j+1}^{m-j} = 1-\alpha^m-\sum_{j=1}^{m-1}\binom{m}{j}F_j(c_1,\ldots,c_j)\alpha_{j+1}^{m-j},$$

leading to

$$\alpha-\alpha^m = \sum_{j=1}^{m-1}\binom{m}{j}F_j(c_1,\ldots,c_j)\alpha_{j+1}^{m-j}.$$

Making use of (5.9), we obtain

$$\frac{\alpha-\alpha^m}{1-\alpha} = \sum_{j=1}^{m-1}\binom{m}{j}\alpha_{j+1}^{m-j} \iff \sum_{j=1}^{m-1}\alpha^j = \sum_{j=1}^{m-2}\binom{m}{j}\alpha_{j+1}^{m-j} + m\alpha_m,$$

hence, finally,

$$\alpha_m = \frac{1}{m}\left[\sum_{j=1}^{m-1}\alpha^j - \sum_{j=1}^{m-2}\binom{m}{j}\alpha_{j+1}^{m-j}\right], \quad m \geq 2. \tag{5.10}$$

For example, it holds $\alpha_2 = \alpha/2, \alpha_3 = \alpha/3 + \alpha^2/12, \alpha_4 = \alpha/4 + \alpha^2/12 + \alpha^3/24 - \alpha^4/96$. These critical values have been derived by Rom (1990).

Corollary 5.2 (Exact FWER-controlling step-up test under independence). *Consider a multiple test problem* $(\mathcal{X}, \mathcal{F}, \mathcal{P}, \mathcal{H}_m)$ *and assume stochastically independent marginal p-values* $p_1, \ldots, p_m$. *Let* $p_{1:m}, \ldots, p_{m:m}$ *denote the ordered p-values and* $H_{1:m}, \ldots, H_{m:m}$ *the correspondingly sorted null hypotheses. Let* $\alpha = \alpha_1 \geq \cdots \geq \alpha_m$ *denote the critical values defined by* (5.10). *Then, the step-up test rejecting exactly* $H_{1:m}, \ldots, H_{m^*:m}$, *where*

$$m^* = \max\{i \in I : p_{i:m} \leq \alpha_{m-i+1}\},$$

controls the FWER at level α, *and it exhausts the FWER level if p-values are exactly uniform under null hypotheses.*

Proof. Exact FWER control of this step-up test follows immediately from the construction of $\alpha_1, \ldots, \alpha_m$ and by considering each possible value of m_0 separately. □

We observe that $\alpha/i \leq \alpha_i \leq 1-(1-\alpha)^{1/i}$ for all $i \in I$. This illustrates the interrelation of the structure of the test procedure, the size of the critical values, and the structural assumptions regarding the joint distribution of p-values (by comparing with Bonferroni-Holm and Šidák-Holm).

5.4 FDR-Controlling Step-Up Tests

The by far most popular FDR-controlling multiple test procedure is the linear step-up test φ^{LSU} (say), considered in the pioneering article by Benjamini and Hochberg (1995). Sometimes it is even referred to as *the* FDR procedure. The test φ^{LSU} is a step-up test with Simes' critical values.

Definition 5.6. Denote by $p_{1:m} \leq p_{2:m} \leq \ldots \leq p_{m:m}$ the ordered p-values for a multiple test problem $(\mathcal{X}, \mathcal{F}, (\mathbb{P}_\vartheta)_{\vartheta \in \Theta}, \mathcal{H}_m = \{H_i, i \in I = \{1, \ldots, m\}\})$ and by $H_{1:m}, \ldots, H_{m:m}$ the re-ordered null hypotheses in $\mathcal{H}_m$, according to the ordering of the p-values. Then, the linear step-up test φ^{LSU} rejects exactly the hypotheses $H_{1:m}, \ldots, H_{k:m}$, where

$$k = \max\{i \in I : p_{i:m} \leq i\alpha/m\}. \tag{5.11}$$

If the maximum in (5.11) does not exist, then no hypothesis is rejected.

The linear step-up test controls the FDR under the PRDS assumption regarding the joint distribution of $p_1, \ldots, p_m$, including cases with independent p-values. More precisely, the following theorem characterizes FDR control of φ^{LSU}.

Theorem 5.3 (Finner et al. (2009)). *Consider the following assumptions.*

(D1) $\forall \vartheta \in \Theta : \forall j \in I : \forall i \in I_0(\vartheta) : \mathbb{P}_\vartheta(R_m \geq j | p_i \leq t)$ *is non-increasing in* $t \in (0, \alpha_{j:m}]$.
(D2) $\forall \vartheta \in \Theta : \quad \forall i \in I_0(\vartheta) : p_i \sim UNI[0, 1]$.
(I1) $\forall \vartheta \in \Theta :$ *The p-values* $(p_i(X) : i \in I_0(\vartheta))$*, are iid.*
(I2) $\forall \vartheta \in \Theta$*: The random vectors* $(p_i(X) : i \in I_0(\vartheta))$ *and* $(p_i(X) : i \in I_1(\vartheta))$ *are stochastically independent.*

Then, the following two assertions hold true.

$$Under(D1), \forall \vartheta \in \Theta : \quad FDR_\vartheta(\varphi^{LSU}) \leq \frac{m_0(\vartheta)}{m}\alpha. \tag{5.12}$$

$$Under(D2) - (I2), \forall \vartheta \in \Theta : \quad FDR_\vartheta(\varphi^{LSU}) = \frac{m_0(\vartheta)}{m}\alpha. \tag{5.13}$$

Theorem 5.3 implies that φ^{LSU} controls the FDR under PRDS, see (5.12). Another remarkable property of φ^{LSU} is that its FDR under (D2) - (I2) depends on ϑ only via $m_0 = m_0(\vartheta)$, see (5.13). The latter fact suggests a data-adaptive modification of φ^{LSU}, leading to the step-up test φ^{STS} (say), introduced by Storey et al. (2004).

Definition 5.7. Denote the ecdf of the p-values $p_1, \ldots, p_m$ by $\hat{F}_m$ and consider the following estimator for the proportion $\pi_0 = m_0/m$ of true hypotheses.

$$\hat{\pi}_0^{\mathrm{STS}} \equiv \hat{\pi}_0^{\mathrm{STS}}(\lambda) = \frac{1 - \hat{F}_m(\lambda) + 1/m}{1 - \lambda}, \quad \lambda \in [0, 1).$$

Then, the data-adaptive step-up test φ^{STS} is given by replacing α by $\alpha/\hat{\pi}_0^{\mathrm{STS}}$ in the definition of φ^{LSU}.

Notice the similarity of $\hat{\pi}_0^{\mathrm{STS}}$ and the Schweder-Spjøtvoll estimator $\hat{\pi}_0$ defined in (3.2). In fact, the extra term $1/m$ in the numerator of $\hat{\pi}_0^{\mathrm{STS}}$ can be regarded as a finite-sample adjustment of $\hat{\pi}_0$ and becomes negligible for large m.

Theorem 5.4 (Storey et al. (2004)).

(a) Under (I1) - (I2) from Theorem 5.3, it holds

$$\sup_{\vartheta \in \Theta} FDR_\vartheta(\varphi^{STS}) \leq (1 - \lambda^{m_0})\alpha \leq \alpha,$$

if hypotheses with p-values larger than λ are removed from the set of rejected hypotheses.

(b) Under weak dependency and assuming that $\lim_{m \to \infty} m_0/m$ *exists, φ^{STS} controls the FDR asymptotically (as $m \to \infty$).*

For the case of arbitrary dependency among p-values, the Benjamini-Yekutieli step-up test has been derived according to the following general construction method.

1. Assume there exists a multiple test procedure φ that can be calibrated such that it controls the FDR at level α over some parameter space Θ^*, i. e., $\sup_{\vartheta \in \Theta^*} \mathrm{FDR}_\vartheta(\varphi) \leq \alpha$.
2. Derive a bound for the FDR of φ over the actual parameter space of interest, i. e., find a constant α' such that $\sup_{\vartheta \in \Theta} \mathrm{FDR}_\vartheta(\varphi) \leq \alpha'$.
3. If there exists an invertable function $h : [0, 1] \to [0, 1]$ such that $\alpha' = h(\alpha)$ and if h does not depend on unknown parameters, then substitute α by $h^{-1}(\alpha)$ in the calibration of φ.

Theorem 5.5 (Benjamini and Yekutieli (2001)). *Let φ^{LSU} denote the linear step-up test considered by Benjamini and Hochberg (1995). Then, for any dependency structure among $p_1, \ldots, p_m$, it holds*

$$\forall \vartheta \in \Theta : FDR_\vartheta(\varphi^{LSU}) \leq \frac{m_0(\vartheta)}{m}\alpha \sum_{j=1}^{m} \frac{1}{j} \leq \alpha \sum_{j=1}^{m} \frac{1}{j}.$$

Hence, defining the function h by $h(\alpha) = \alpha \sum_{j=1}^{m} j^{-1}$, the test φ^{LSU} with α replaced by $h^{-1}(\alpha) = \alpha / \sum_{j=1}^{m} j^{-1}$ controls the FDR under arbitrary dependency among $p_1, \ldots, p_m$. We denote this modified version of φ^{LSU} by φ^{BY}.

Guo and Rao (2008) have shown that there indeed exists a multivariate distribution of p-values $p_1, \ldots, p_m$ such that φ^{BY} fully exhausts the FDR level α. Hence, in the class of step-up tests with fixed critical values, φ^{BY} cannot be improved uniformly if the dependency structure is completely unknown.

A different way to construct FDR-controlling step-up tests has been proposed by Blanchard and Roquain (2008).

Theorem 5.6 (Blanchard and Roquain (2008)). *Let φ^{ν} denote a step-up test with critical values of the form $\alpha_{j:m} = \alpha\beta(j)/m$, where*

$$\beta(j) \equiv \beta_{\nu}(j) = \int_0^j x d\nu(x) \tag{5.14}$$

for an arbitrarily chosen probability distribution ν on $(0, \infty)$. Then, φ^{ν} controls the FDR under arbitrary dependency among $p_1, \ldots, p_m$.

The class of step-up tests defined by Theorem 5.6 contains φ^{BY} as a special member. To see this, consider the probability distribution ν^{BY} (say) which is supported on $\{1, \ldots, m\}$ and defined by $\nu^{BY}(\{k\}) = (k \sum_{\ell=1}^{m} \ell^{-1})^{-1}, k \in \{1, \ldots, m\}$. As shown by Blanchard and Roquain (2008), also other FDR-controlling step-up tests proposed in the literature can be obtained as special cases of φ^{ν} (for appropriately chosen ν).

5.5 FDR-Controlling Step-Up-Down Tests

Since the set of hypotheses which are rejected by a step-up-down test φ^{κ_1} of order κ_1 is contained in the set of hypotheses which are rejected by φ^{κ_2} whenever $\kappa_2 > \kappa_1$ and the critical values $\alpha_{1:m}, \ldots, \alpha_{m:m}$ are kept fixed (see Lemma 3.1), it is natural to ask if FDR control of φ^{κ_2} implies FDR control of φ^{κ_1}. In particular, by setting $\kappa_2 = m$ and making use of the step-up tests derived in the previous section, the question arises if the critical values employed in these step-up tests can also be utilized to control the FDR with a corresponding step-up-down test. Sufficient conditions for a positive answer to this question have been provided by Zeisel et al. (2011).

Theorem 5.7 (Theorem 4.1 by Zeisel et al. (2011)). *Under the scope of a two-class mixture model, assume that all m p-values are stochastically independent. If F_1 is a concave cdf. and φ_1 and φ_2 are two multiple tests such that the set of hypotheses which are rejected by φ_1 is contained in the set of hypotheses which are rejected by φ_2 for any realization of the vector of p-values, then the FDR of φ_1 is upper-bounded by the FDR of φ_2.*

Hence, under the assumption of joint independence of the p-values in a two-class mixture model, FDR-controlling step-up-down tests can be derived from the procedures investigated in Sect. 5.4, provided F_1 is concave. In the remainder of this section, we discuss a different class of FDR-controlling step-up-down tests which make use of FDR monotonicity for a fixed, given multiple test with respect to the distribution of p-values under alternatives. The starting point is the following monotonicity theorem for step-up tests.

Theorem 5.8 (Benjamini and Yekutieli (2001)). *Let conditions (D1) - (I2) from Theorem 5.3 be fulfilled. Then, an SU-procedure φ with critical values $\alpha_{1:m} \leq \cdots \leq \alpha_{m:m}$ has the following properties.*

(a) If the ratio $\alpha_{i:m}/i$ is increasing in i, as $(p_i : i \in I_1)$ increases stochastically, the FDR of φ decreases.
(b) If the ratio $\alpha_{i:m}/i$ is decreasing in i, as $(p_i : i \in I_1)$ increases stochastically, the FDR of φ increases.

Hence, if p-values are independent, φ is a step-up test and the associated critical values $\alpha_{1:m} \leq \cdots \leq \alpha_{m:m}$ fulfill the important condition

$$\alpha_{i:m}/i \text{ is increasing in } i, \tag{5.15}$$

then Dirac-uniform configurations are least favourable for the FDR of φ, provided that Dirac-uniform configurations are elements of the statistical model under consideration or limiting elements thereof. Often, extreme values of the parameter of a statistical model under alternatives lead to p-values which are arbitrarily close to zero such that the latter reasoning applies. Sets of critical values fulfilling condition (5.15) are called *feasible critical values* by Finner et al. (2012). Theorem 5.8 suggests the following method for constructing an FDR-exhausting step-up test.

Algorithm 5.1

(i) For every $1 \leq m_0 \leq m$, consider the Dirac-uniform configuration $DU_{m_0,m}$ and the equation $FDR_{m_0,m}(\alpha_{1:m}, \ldots, \alpha_{m:m}) = \alpha$, where $\alpha_{1:m}, \ldots, \alpha_{m:m}$ are the critical values that should be optimized for usage in a step-up test. Recall that the joint point mass function of V_m and R_m can exactly be computed under $DU_{m_0,m}$.
(ii) Solve the resulting system of m equations for the m unknowns $\alpha_{1:m}, \ldots, \alpha_{m:m}$.
(iii) Check if the solution is feasible in the sense that the optimized values $\alpha_{1:m}, \ldots, \alpha_{m:m}$ fulfill condition (5.15).

This strategy has been pursued by Kwong and Wong (2002) and refined by Finner et al. (2012). However, it turns out that the formal solution of the system of equation is often not feasible. For instance, if $\alpha = 0.05$, solutions are only feasible for $m \leq 6$. Hence, the formal solution typically has to be modified to fulfill (5.15).

However, Finner et al. (2009) have shown that the general reasoning of Algorithm 5.1 applies in an asymptotic sense when $m \to \infty$. To this end, it is important to notice that many stepwise rejective multiple tests can be described in terms of critical value functions and rejection curves.

Definition 5.8. We call a non-decreasing, continuous function $r : [0, 1] \to [0, 1]$ a rejection curve. Its generalized inverse $\rho = r^{-1}$ is called the corresponding critical value curve.

Using these definitions, we observe that critical values may be defined by $\alpha_{i:m} = \rho(i/m)$, $1 \leq i \leq m$. Feasibility of these critical values can be assured by requiring that

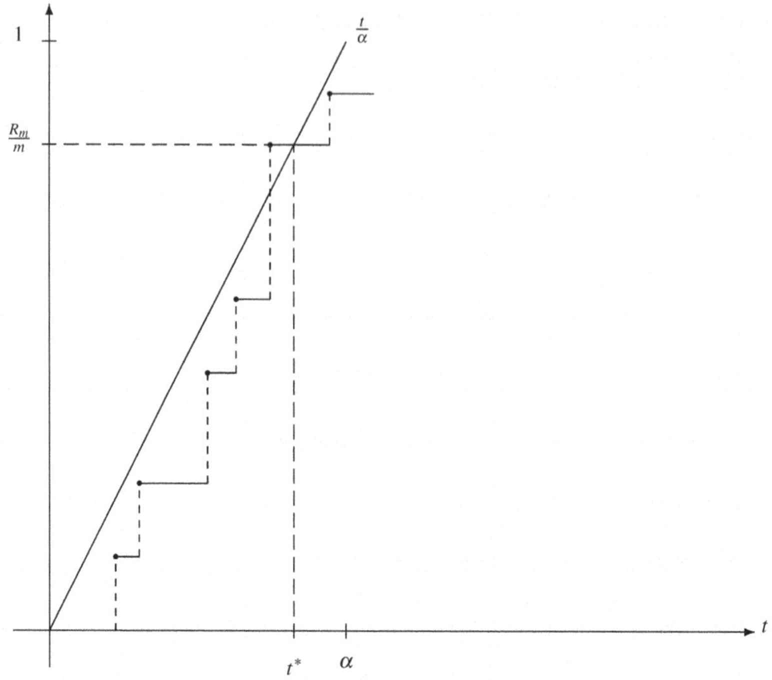

Fig. 5.1 Graphical illustration of φ^{LSU} in terms of Simes' line

$$q(t) = \rho(t)/t \text{ is non-decreasing in } t \in [0, 1]. \tag{5.16}$$

In general, r and ρ depend on m. However, in asymptotic considerations we will assume that they are fixed, given objects. The following important result establishes the link between ordered p-values, the ecdf. $\hat{F}_m$ of p-values, critical values and the rejection curve formally.

Lemma 5.6 (Sen (1999)).

$$p_{i:m} \leq \alpha_{i:m} \textit{ if and only if } \hat{F}_m(p_{i:m}) \geq r(p_{i:m}).$$

Hence, the rejection threshold t^* (say) for the p-values can equivalently be expressed as the abscissa of a crossing point of $\hat{F}_m$ and r, yielding a useful graphical illustration to see how a particular step-up-down test works, namely, to draw $\hat{F}_m$ and r in one graph. Furthermore, the ordinate of this crossing point equals the proportion R_m/m of rejected hypotheses. For instance, the linear step-up test φ^{LSU} can equivalently be defined in terms of the rejection curve $r(t) = t/\alpha$, called Simes' line, see Fig. 5.1. Notice that step-up means that the largest crossing point determines t^*.

More generally, the choice of the parameter κ of a step-up-down test corresponds to choosing a point $\lambda \in [0, 1]$ in the vicinity of which the crossing point of $\hat{F}_m$ and r is determined to derive t^*. More formally, we obtain the following equivalent formulation of a step-up-down test in terms of this tuning parameter λ.

Lemma 5.7. *The rejection threshold t^* for p-values generated by a step-up-down test with tuning parameter $\lambda \in [0, 1]$ is given by*

$$t^* = \begin{cases} \inf\{p_i > \lambda : \hat{F}_m(p_i) < r(p_i)\}, & \text{if } \hat{F}_m(\lambda) \geq r(\lambda), \quad (SD - branch) \\ \sup\{p_i < \lambda : \hat{F}_m(p_i) \geq r(p_i)\}, & \text{if } \hat{F}_m(\lambda) < r(\lambda), \quad (SU - branch) \end{cases}$$

with the additional conventions that all hypotheses are rejected if $\lambda = 1$ (i. e., in the case of a step-up test) and $\hat{F}_m(1) \geq r(1)$ as well as that no hypotheses are rejected if $\lambda = 0$ (i. e., in the case of a step-down test) and $\hat{F}_m(0) < r(0)$.

Of course, the parameter $\kappa \in \{1, \ldots, m\}$ appearing in the original formulation of a step-up-down test (see Definition 3.1) can be translated into the corresponding parameter $\lambda \in [0, 1]$ once the rejection curve is fixed.

In the case that p-values are independent and the rejection curve r is (at least asymptotically) a fixed, given object, the alternative formulation of step-up-down tests provided in Lemma 5.7 is a very helpful tool for the mathematical analysis of step-up-down tests. The reason is that the ecdf $\hat{F}_m$ converges almost surely due to the extended Glivenko-Cantelli theorem.

Theorem 5.9 (Shorack and Wellner (1986), p.105f.). *Let $p_1, \ldots, p_m$ denote stochastically independent p-values, with marginal cdfs $F_1, \ldots, F_m$ respectively. Let $\overline{F} = m^{-1} \sum_{i=1}^m F_i$. Then, it holds*

$$||\hat{F}_m - \overline{F}||_\infty \to 0 \text{ almost surely as } m \to \infty.$$

Corollary 5.3. *Under $DU_{m_0,m}$, assume that $m_0/m \to \pi_0 = 1 - \pi_1 \in (0, 1]$. Then,*

$$\hat{F}_m(t) \to \overline{F}(t) = \pi_1 + \pi_0\, t \text{ uniformly in } t \in [0, 1] \text{ and almost surely as } m \to \infty. \tag{5.17}$$

Based on these considerations, a (heuristic) asymptotic analogue of Algorithm 5.1 consists of finding an "asymptotically optimal rejection curve" (AORC) r_α (say) such that for any value of $\pi_0 > \alpha$, an appropriate crossing point of r_α and $\overline{F}$ from (5.17) is such that $\lim_{m\to\infty} \mathrm{FDR}_{m_0,m}(\varphi) = \alpha$, where φ is a stepwise rejective multiple test based on r_α. We recall that the ordinate of such a crossing point (the abscissa of which equals the p-value threshold t^*) equals the proportion R_m/m of rejected hypotheses. Since under $DU_{m_0,m}$, $R_m = V_m + m_1$ almost surely, we obtain that

$$\lim_{m\to\infty} \mathrm{FDR}_{m_0,m}(\varphi) = \frac{\pi_0 t^*}{\pi_1 + \pi_0 t^*}.$$

Thus, the optimal threshold $t^*(\pi_0)$ fulfills the relationship

$$\frac{\pi_0 t^*(\pi_0)}{\pi_1 + \pi_0 t^*(\pi_0)} = \alpha \quad \Leftrightarrow \quad t^*(\pi_0) = \frac{\alpha \pi_1}{\pi_0(1-\alpha)}$$

and the AORC r_α is given by

$$r_\alpha(t^*(\pi_0)) = \pi_1 + \pi_0 t^*(\pi_0) \Longleftrightarrow r_\alpha(u) = \frac{u}{\alpha + (1-\alpha)u}, \quad u \in [0, 1],$$

by substituting $u = t^*(\pi_0) = \alpha(1-\pi_0)/(\pi_0(1-\alpha))$. Since the asymptotically optimal threshold $t^*(\pi_0)$ is generated by the AORC automatically for every $\pi_0 > \alpha$, procedures based on r_α are called implicitly adaptive, in contrast to (explicitly) adaptive procedures like φ^{STS} which pre-estimate π_0.

The critical value curve associated with r_α is given by

$$r_\alpha^{-1}(t) = \frac{\alpha t}{1-(1-\alpha)t} = 1 - r_\alpha(1-t), \quad t \in [0, 1]$$

leading to the AORC-induced critical values

$$\alpha_{i:m} = r_\alpha^{-1}(i/m) = \frac{\frac{i}{m}\alpha}{1-\frac{i}{m}(1-\alpha)} = \frac{i\alpha}{m-i(1-\alpha)}, \quad 1 \le i \le m.$$

These critical values are feasible for all $\alpha \in (0, 1)$. However, unfortunately, $\alpha_{m:m} \equiv 1$ such that a step-up test based on the AORC would always reject all hypotheses, hence not controlling the FDR. Actually, for any $\pi_0 \in (\alpha, 1]$, there are exactly two crossing points of r_α and $\overline{F}$ from (5.17), the larger of which is equal to $(1, 1)$, cf. Fig. 5.2.

Any AORC-based procedure that excludes $(1, 1)$ as the crossing point determining the rejection threshold t^* asymptotically controls the FDR at level α under joint independence of all p-values. In particular, the following result holds true for step-up-down tests induced by r_α.

Theorem 5.10 (Finner et al. (2009)). *Assume the distributional assumptions (D2)–(I2) from Theorem 5.3 hold true. Then, any SUD-procedure with parameter $\lambda \in [0, 1)$ based on r_α asymptotically controls the FDR at level α.*

Notice that the proof of Theorem 5.10 can not straightforwardly make use of Theorem 5.8, because the latter theorem is exclusively dealing with step-up tests. As shown recently by Blanchard et al. (2014), Dirac-uniform configurations are in general not least favorable for step-up-down tests with parameter $\lambda \in [0, 1)$ for finite m, but the difference between the FDR under the least favorable configurations and the FDR under Dirac-uniform configurations asymptotically vanishes for any value of $\pi_0 \in [0, 1]$ if $\lambda > 0$ and for any value of $\pi_0 \in [0, 1)$ if $\lambda = 0$.

Moreover, the maximum FDR under Dirac-uniform configurations of an SUD-procedure with parameter $\lambda \in [0, 1)$ based on r_α often approaches α from above,

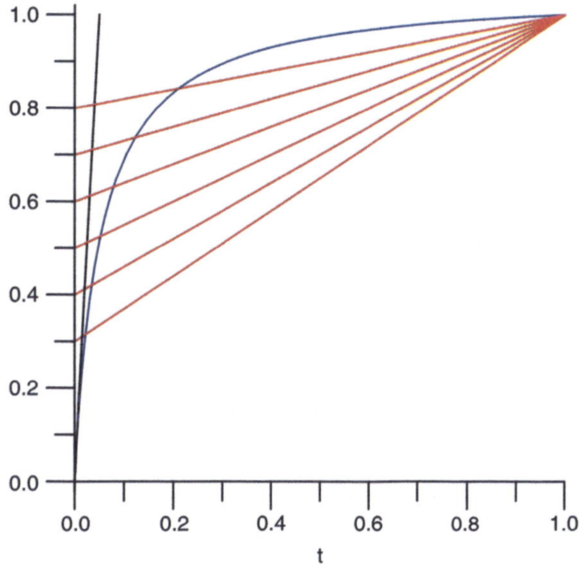

Fig. 5.2 Simes' line (*black*), AORC (*blue*) and $\overline{F}$ from (5.17) for different values of π_0 (*red*), where $\alpha = 0.05$

meaning that the FDR level is violated for finite systems of hypotheses, at least if the Dirac-uniform configurations belong to the considered model class or are limiting elements thereof, see Fig. 5.3 for an illustration.

Hence, an adjustment of $\alpha_{1:m}, \ldots, \alpha_{m:m}$ is necessary if m is small or moderate. Finner et al. (2012) investigate several possible adjustment methods. Other modifications of AORC-based critical values are discussed by Blanchard and Roquain (2009). Since Dirac-uniform configurations are in general not least favorable for step-up-down tests with $\kappa < m$ (see the counterexamples by Blanchard et al. (2014)), some other FDR bounds for step-up-down tests are required to calibrate critical values for the case of finite m. Such bounds have been derived by Finner et al. (2009) and Finner et al. (2012).

Theorem 5.11 (Finner et al. (2012)). *A ϑ-free upper bound for the FDR of a step-up-down test φ^κ with parameter $\kappa \in \{1, \ldots, m\}$ based on critical values $\alpha_{1:m}, \ldots, \alpha_{m:m}$ is for fixed $1 \leq m_0 \leq m$ given by*

$$b(m, m_0|\kappa) = \frac{m_0}{m}\mathbb{E}_{m_0-1,m}[q(R_m/m)]$$
$$= m_0 \sum_{j=1}^{m_0} \frac{\alpha_{m_1+j:m}}{m_1 + j}\mathbb{P}_{m_0-1,m}(V_m = j - 1). \tag{5.18}$$

Hence, the FDR of φ^κ is bounded by

$$b(m|\kappa) = \max_{1 \leq m_0 \leq m} b(m, m_0|\kappa). \tag{5.19}$$

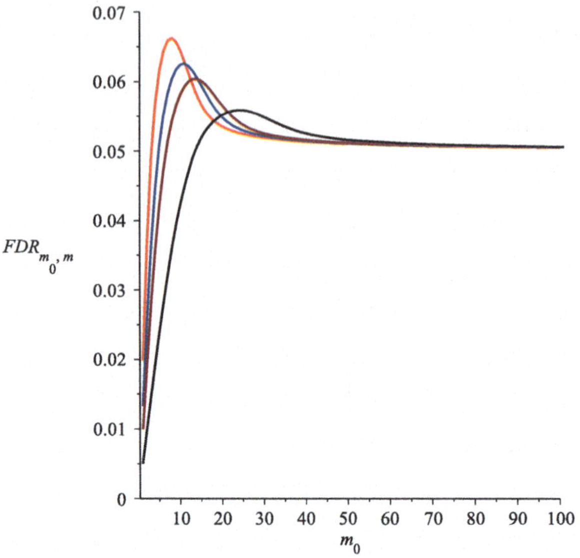

Fig. 5.3 FDR under Dirac-uniform configurations of an SUD-procedure with parameter $\lambda = 1/2$ based on r_α, where $\alpha = 0.05$, as a function of m_0. The four different curves correspond to $m = 50$ (*red*), $m = 75$ (*blue*), $m = 100$ (*brown*), and $m = 200$ (*black*). Discrete FDR values have been interpolated and FDR values for $m_0 > 100$ in the case of $m = 200$ have been omitted, for a better visualization

In the case of step-up, that is $\kappa = m$, *the bound* $b(m, m_0|\kappa)$ *equals* $FDR_{m_0,m}(\varphi^m)$, *which results in the alternative formula*

$$b(m, m_0|\kappa = m) = \sum_{j=1}^{m_0} \frac{j}{m_1 + j} \mathbb{P}_{m_0,m}(V_m = j), \tag{5.20}$$

and it even holds equality in every summand of (5.18) *and* (5.20), *yielding the nice recursive formula*

$$\forall 1 \leq j \leq m_0 : \mathbb{P}_{m_0,m}(V_m = j) = \frac{m_0}{j} \alpha_{m_1+j:m} \mathbb{P}_{m_0-1,m}(V_m = j - 1). \tag{5.21}$$

Theorem 5.11 provides a convenient way to adjust critical values for strict FDR control for finite m. All that has to be done is checking that $b(m|\kappa)$ does not exceed α. For the special case of AORC-induced critical values, the following modification can easily be implemented for practical usage. One tries to find a constant $\beta_m > 0$ such that the set of critical values

$$\alpha_{i:m} = \frac{i\alpha}{m + \beta_m - i(1 - \alpha)}, \quad 1 \leq i \leq m, \tag{5.22}$$

which are always feasible, yield that $b(m|\kappa) \leq \alpha$. A nice analytical result in this direction has been derived by Gavrilov et al. (2009). They show that for step-down, i. e., in the case of $\kappa = 1$, the choice $\beta_m \equiv 1$ leads to strict FDR control for any number m of hypotheses. For $\kappa \in \{2, \ldots, m\}$, optimal values for β_m are tabulated or computer programs exist to compute β_m, cf. Finner et al. (2012).

Acknowledgments Parts of Sects. 5.2 and 5.3 were inspired by material from unpublished lecture notes by Helmut Finner and Iris Pigeot. Sects. 5.4 and 5.5 originated from joint work with Helmut Finner, Markus Roters, and Veronika Gontscharuk. I thank Gilles Blanchard and Etienne Roquain for working together with me on LFCs for SUD tests under the scope of the PROCOPE 2010 programme of the German Academic Exchange Service (DAAD). Special thanks are due to Mareile Große Ruse for programming the LaTeX code used for Fig. 5.1.

References

Benjamini Y, Hochberg Y (1995) Controlling the false discovery rate: A practical and powerful approach to multiple testing. J R Stat Soc Ser B Stat Methodol 57(1):289–300

Benjamini Y, Yekutieli D (2001) The control of the false discovery rate in multiple testing under dependency. Ann Stat 29(4):1165–1188

Blanchard G, Roquain E (2008) Two simple sufficient conditions for FDR control. Electron J Statist 2:963–992

Blanchard G, Roquain E (2009) Adaptive false discovery rate control under independence and dependence. J Mach Learn Res 10:2837–2871

Blanchard G, Dickhaus T, Roquain E, Villers F (2014) On least favorable configurations for step-up-down tests. Statistica Sinica 24(1):1–23

Block HW, Savits TH, Wang J, Sarkar SK (2013) The multivariate-t distribution and the Simes inequality. Stat Probab Lett 83(1):227–232. doi:10.1016/j.spl.2012.08.013

Dalal S, Mallows C (1992) Buying with exact confidence. Ann Appl Probab 2(3):752–765. doi:10.1214/aoap/1177005658

Finner H, Dickhaus T, Roters M (2009) On the false discovery rate and an asymptotically optimal rejection curve. Ann Stat 37(2):596–618. doi:10.1214/07-AOS569

Finner H, Gontscharuk V, Dickhaus T (2012) False discovery rate control of step-up-down tests with special emphasis on the asymptotically optimal rejection curve. Scand J Stat 39:382–397

Gavrilov Y, Benjamini Y, Sarkar SK (2009) An adaptive step-down procedure with proven FDR control under independence. Ann Stat 37(2):619–629. doi:10.1214/07-AOS586

Guilbaud O (2008) Simultaneous confidence regions corresponding to Holm's step-down procedure and other closed-testing procedures. Biom J 50(5):678–692

Guilbaud O (2012) Simultaneous confidence regions for closed tests, including Holm-, Hochberg-, and Hommel-related procedures. Biom J 54(3):317–342

Guilbaud O, Karlsson P (2011) Confidence regions for Bonferroni-based closed tests extended to more general closed tests. J Biopharm Stat 21(4):682–707

Guo W, Rao M (2008) On control of the false discovery rate under no assumption of dependency. J Stat Plann Infer 138(10):3176–3188. doi:10.1016/j.jspi.2008.01.003

Hochberg Y (1988) A sharper Bonferroni procedure for multiple tests of significance. Biometrika 75(4):800–802. doi:10.1093/biomet/75.4.800

Holm S (1979) A simple sequentially rejective multiple test procedure. Scand J Stat Theory Appl 6:65–70

Holm SA (1977) Sequentially rejective multiple test procedures. Statistical Research Report No. 1977–1. Institute of Mathematics and Statistics, University of Umeå.

Hommel G (1988) A stagewise rejective multiple test procedure based on a modified Bonferroni test. Biometrika 75(2):383–386. doi:10.1093/biomet/75.2.383

Hommel G, Bretz F, Maurer W (2007) Powerful short-cuts for multiple testing procedures with special reference to gatekeeping strategies. Stat Med 26(22):4063–4073

Hu T, Chen J, Xie C (2006) Regression dependence in latent variable models. Probab Eng Inf Sci 20(2):363–379. doi:10.1017/S0269964806060220

Karlin S, Rinott Y (1980) Classes of orderings of measures and related correlation inequalities. I. Multivariate totally positive distributions. J Multivariate Anal 10:467–498

Kwong KS, Wong EH (2002) A more powerful step-up procedure for controlling the false discovery rate under independence. Stat Probab Lett 56(2):217–225

Rom DM (1990) A sequentially rejective test procedure based on a modified Bonferroni inequality. Biometrika 77:663–665

Sarkar SK (1998) Some probability inequalities for ordered MTP_2 random variables: a proof of the Simes conjecture. Ann Stat 26(2):494–504. doi:10.1214/aos/1028144846

Sarkar SK (2002) Some results on false discovery rate in stepwise multiple testing procedures. Ann Stat 30(1):239–257

Sarkar SK, Chang CK (1997) The Simes method for multiple hypothesis testing with positively dependent test statistics. J Am Stat Assoc 92(440):1601–1608. doi:10.2307/2965431

Sen PK (1999) Some remarks on Simes-type multiple test of significance. J Statist Plann Infer 82:139–145

Shorack GR, Wellner JA (1986) Empirical processes with applications to statistics. Wiley series in probability and mathematical statistics. John Wiley & Sons Inc., New York

Simes RJ (1986) An improved Bonferroni procedure for multiple tests of significance. Biometrika 73:751–754

Storey JD, Taylor JE, Siegmund D (2004) Strong control, conservative point estimation and simultaneous conservative consistency of false discovery rates: a unified approach. J R Stat Soc Ser B Stat Methodol 66(1):187–205

Strassburger K, Bretz F (2008) Compatible simultaneous lower confidence bounds for the Holm procedure and other Bonferroni-based closed tests. Stat Med 27(24):4914–4927

Tong Y (1990) The multivariate normal distribution. Springer series in statistics. Springer, New York

Troendle JF, Westfall PH (2011) Permutational multiple testing adjustments with multivariate multiple group data. J Stat Plann Infe 141(6):2021–2029. doi:10.1016/j.jspi.2010.12.012

Zeisel A, Zuk O, Domany E (2011) FDR control with adaptive procedures and FDR monotonicity. Ann Appl Stat 5(2A):943–968

Chapter 6
Multiple Testing and Binary Classification

Abstract We describe connections between multiple testing and binary classification. Under certain sparsity assumptions, classical multiple tests controlling a type I error rate at a fixed level α can, at least asymptotically as the number of classification trials tends to infinity, achieve the optimal (Bayes) classification risk. Under non-sparsity, combinations of type I and type II error rates are discussed as appropriate proxies for the (weighted) misclassification risk, and we provide algorithms for binary classification which are based on multiple testing. The problem of feature selection for binary classification is addressed by the higher criticism criterion, a concept originally introduced for testing the global null hypothesis in a multiple test problem.

Binary classification denotes the problem of assigning random objects to one of exactly two classes. This problem is often addressed by statistical learning techniques, see, for instance, Hastie et al. (2009) and Vapnik (1998) for introductions. Binary classification and multiple testing are related statistical fields. The decision pattern of a multiple test for a family of m hypotheses has the same structure as the output of a binary classifier for m data points to be classified, namely, a vector in $\{0, 1\}^m$ indicating the m binary decisions. Moreover, in both problems typically realizations x_i, $1 \leq i \leq m$, of random vectors X_i with values in $\mathbb{R}^k$ build the basis for the decision rule (the multiple test or the classifier) which is thus chosen according to statistical criteria. In the testing context, x_i has the interpretation of a data sample (or the value of a sufficient statistic) for the i-th individual test, while x_i is referred to as the i-th feature vector to be classified in the classification terminology. On the other hand, usual loss functions for binary classification differ from the ones that are typically utilized in multiple testing.

Definition 6.1. Let $(X_1, Y_1), \ldots, (X_m, Y_m)$ denote stochastically independent and identically distributed random tuples, where X_i takes values in $\mathbb{R}^k$ and Y_i is a binary indicator with values in $\{0, 1\}$, $1 \leq i \leq m$. Let the data-generating process be modeled by a (joint) probability measure $\mathbb{P}$, where some systematic relationship between Y_1 and X_1 is assumed. Namely, the random vectors $X_1, \ldots, X_m$ are assumed

T. Dickhaus, *Simultaneous Statistical Inference*,
DOI: 10.1007/978-3-642-45182-9_6,

to be continuously distributed with class-conditional cdfs given by $F_j(x) = \mathbb{P}(X_i \leq x | Y_i = j)$ for $x \in \mathbb{R}^k, j = 0, 1$ and $i = 1, \ldots, m$. Assume that the "labels" $Y_1, \ldots, Y_m$ can not be observed. Formally, we describe the classification task by the pairs of hypotheses $H_i : Y_i = 0$ versus $K_i : Y_i = 1, 1 \leq i \leq m$.

(a) For a given cost parameter $c \in (0, 1)$ and a rejection region $\Gamma \subset \mathbb{R}^k$, the Bayes risk associated with the action $a_i = \mathbf{1}_\Gamma(x_i)$ is given by

$$R^{(i)}_{\text{Bayes}}(\Gamma) = (1 - c)\mathbb{P}(X_i \in \Gamma, Y_i = 0) + c\mathbb{P}(X_i \notin \Gamma, Y_i = 1). \tag{6.1}$$

Under the additive risk assumption, this entails that the Bayes risk for all m classification tasks together is given by

$$\begin{aligned} R_{\text{Bayes}}(\Gamma) &= \sum_{i=1}^{m} R^{(i)}_{\text{Bayes}}(\Gamma) \\ &= (1 - c)\mathbb{E}[V_m] + c\mathbb{E}[T_m], \end{aligned} \tag{6.2}$$

where the multiple testing error quantities V_m and T_m are as in Table 1.1 and refer to a multiple test with fixed rejection region Γ for every marginal test.

(b) Let a data-dependent classification rule be given by a measurable random mapping $\hat{h}_m : \mathbb{R}^k \to \{0, 1\}$, where we use the observed data $X_1 = x_1, \ldots, X_m = x_m$ to construct the rule $\hat{h}_m$. Then, the transductive and the inductive risk of $\hat{h}_m$, respectively, are given by

$$R^{(T)}(\hat{h}_m) = m^{-1} \sum_{i=1}^{m} \mathbb{P}(\hat{h}_m(X_i) \neq Y_i), \tag{6.3}$$

$$R^{(I)}(\hat{h}_m) = \mathbb{P}(\hat{h}_m(X_{m+1}) \neq Y_{m+1}), \tag{6.4}$$

where the tuple $(X_{m+1}, Y_{m+1}) \sim (X_1, Y_1)$ is stochastically independent of all (X_i, Y_i) for $1 \leq i \leq m$.

Similarly as in the Neyman-Pearson fundamental lemma, a set of best rejection regions $\Gamma = \Gamma_c$ considered in part (a) of Definition 6.1 is given by (see, for instance, Sect. 5.3.3 in Berger 1985)

$$\Gamma_c = \left\{ x \in \mathbb{R}^k : \frac{\pi_0 f_0(x)}{\pi_0 f_0(x) + \pi_1 f_1(x)} \leq c \right\} = \left\{ x \in \mathbb{R}^k : \lambda(x) \geq \frac{\pi_0(1 - c)}{\pi_1 c} \right\}, \tag{6.5}$$

where f_j is the pdf (or likelihood function) corresponding to F_j, $j = 0, 1$, $\lambda(x) = f_1(x)/f_0(x)$ denotes the likelihood ratio, and $\pi_0 = 1 - \pi_1 = \mathbb{P}(Y_1 = 0)$. The interpretation of part (b) of Definition 6.1 is that $\mathbb{P}$ is typically unknown or only partially known in practice and that the data-dependent classifier $\hat{h}_m$ "learns" a rejection region Γ from the observed data. Notice that (at least for non-extreme values of c) the classification risk measures introduced in Definition 6.1 do not imply a fixed bound on a multiple type I error rate like the FWER or the FDR, but are of different type in the sense that type I and type II errors are treated (more or less) symmetrically. Weighting of type I and type II errors (i. e., misclassifying a member of the "0"-class as "1" and vice versa) is possible by choosing c appropriately. On the other hand, the data themselves implicitly also induce a weighting, namely by the relative frequencies of the true, but unobserved labels ($m_0 = |\{1 \leq i \leq m : y_i = 0\}|$ and $m_1 = m - m_0$).

6.1 Binary Classification Under Sparsity

From the preceding discussion, it becomes clear that multiple tests controlling a type I error rate like the FWER or the FDR at a fixed significance level are in general not good classifiers, because they treat null hypotheses and alternatives asymmetrically in the underlying risk criterion. Remarkable exceptions are sparse cases where class probabilities are highly unbalanced. Under sparsity, multiple tests can, at least asymptotically ($m \to \infty$), achieve optimal classification risks. As noted by Neuvial and Roquain (2012), the optimal rejection region Γ_c in (6.5) simplifies to a threshold for the data point x_i itself if $k = 1$ and the likelihood ratio λ is increasing in its argument x. The label $\hat{y}_i = 1$ is chosen if x_i exceeds a certain value. If a model for F_0 is available, this rule can equivalently be formalized by deciding $\hat{y}_i = 1$ if the p-value $p_i(x) = 1 - F_0(x_i)$ falls below the corresponding threshold on the p-value scale. This connects the theory of binary classification with that of p-value based multiple hypotheses testing that we have considered in the previous chapters of the present work.

It seems that Abramovich et al. (2006) were the first to realize that the linear step-up test φ^{LSU} from Definition 5.6, which has originally been developed for FDR control under independence, has remarkable properties with respect to a broad range of risk measures under sparsity assumptions, meaning that m_1 is small. While Abramovich et al. (2006) considered the particular problem of estimation under ℓ_r loss in high dimensions by employing thresholding estimators, their findings have also been the basis for studying classification risk properties of φ^{LSU} under sparsity. Bogdan et al. (2011) defined the concept of "Asymptotic Bayes optimality under sparsity" (ABOS) in a normal scale mixture model.

Definition 6.2. (Bogdan et al. (2011)). Under the assumptions of Definition 6.1, let $k = 1$. For the distribution of the independent observables $X_i : 1 \leq i \leq m$, consider the Bayesian model

$$X_i|\mu_i \sim \mathcal{N}(\mu_i, \sigma_\varepsilon^2),$$
$$\mu_i \sim \pi_0\mathcal{N}(0, \sigma_0^2) + \pi_1\mathcal{N}(0, \sigma_0^2 + \tau^2),$$

where the μ_i, $1 \leq i \leq m$, are stochastically independent and σ_0^2 may be equal to zero. Hence, marginally, $X_i \sim \pi_0 \mathcal{N}(0, \sigma^2) + \pi_1 \mathcal{N}(0, \sigma^2 + \tau^2)$, with $\sigma^2 = \sigma_\varepsilon^2 + \sigma_0^2$. Assuming σ^2 to be known, the optimal rejection region Γ_c from (6.5) is such that $\hat{y}_i = 1$ if and only if $x_i^2/\sigma^2 \geq K^2$, where the cutoff K^2 is given by

$$K^2 = (1 + 1/u)\{\log(v) + \log(1 + 1/u)\}, \tag{6.6}$$

with $u = (\tau/\sigma)^2$ and $v = u(\pi_0/\pi_1)^2\delta^2$, $\delta = (1 - c)/c$. Denote the Bayes risk $R_{\text{Bayes}}(\Gamma)$, evaluated at this best rejection region, by R_{opt}. In practice, K^2 can typically not be computed exactly, because π_0 and/or τ^2 may be unknown. For a given multiple test procedure φ operating on $x_1, \ldots, x_m$, let $R_{\text{Bayes}}(\varphi)$ denote the risk functional defined in (6.2), with V_m and T_m now referring to φ. Assume that the model is such that

$$\pi_1 = \pi_1(m) \to 0,\ u = u(m) \to \infty,\ v = v(m) \to \infty, \text{ and } \log(v)/u \to C \in (0, \infty), \tag{6.7}$$

as $m \to \infty$ (where convergence or divergence, respectively, may be along a subsequence indexed by $t = 1, 2, \ldots$). Notice that the dependence of v on m may imply that c depends on m, too. Then, φ is called asymptotically Bayes optimal under sparsity (ABOS), if

$$\frac{R_{\text{Bayes}}(\varphi)}{R_{opt}} \to 1, \quad t \to \infty. \tag{6.8}$$

It is clear that, under the conditions given in (6.7), eventually (for large m) the type I error component of the Bayes risk will dominate the type II error component, due to sparsity. Consequently, it turns out that classical multiple tests which are targeted towards type I error control are ABOS in the sense of Definition 6.2, at least for particular parameter configurations.

Theorem 6.1 (Bogdan et al. (2011)). *Under the model assumptions from Definition 6.2, the following assertions hold true.*

(a) *Consider the Bonferroni test $\varphi^{Bonf} = (\varphi_i^{Bonf} : 1 \leq i \leq m)$ (cf. Example 3.1) operating on $x_1, \ldots, x_m$, the FWER level $\alpha = \alpha(m)$ of which fulfills $\alpha(m) \to \alpha_\infty \in [0, 1)$ such that $\alpha(m)/(1 - \alpha(m)) \propto (\delta\sqrt{u})^{-1}$. Then, φ^{Bonf} is ABOS if $\pi_1(m) \propto m^{-1}$. The condition imposed on α_m means that the Bayesian FDR (see Efron and Tibshirani (2002)) of φ^{Bonf} is proportional to α_m.*
(b) *Consider the linear step-up test φ^{LSU} from Definition 5.6 operating on $x_1, \ldots, x_m$, the FDR level $\alpha = \alpha(m)$ of which fulfills $\alpha(m) \to \alpha_\infty \in [0, 1)$ such that $\alpha(m)/(1 - \alpha(m)) \propto (\delta\sqrt{u})^{-1}$. Then, φ^{LSU} is ABOS whenever $\pi_1(m) \to 0$ such that $m\pi_1(m) \to s \in (0, \infty]$ as $m \to \infty$. In this sense, φ^{LSU} adapts to the unknown degree of sparsity in the data.*

Neuvial and Roquain (2012) generalized the findings of Bogdan et al. (2011) concerning φ^{LSU} to a broader class of distributions of X_1. Namely, they assumed that the (conditional) distribution of X_1 given $Y_1 = 0$ belongs to the parametric family considered by Subbotin (1923).

Definition 6.3. For a given shape parameter $\zeta \geq 1$, the distribution with Lebesgue density f_ζ, given by

$$f_\zeta(x) = \exp(-|x|^\zeta/\zeta)\{2\Gamma(1/\zeta)\zeta^{1/\zeta-1}\}^{-1}, \quad x \in \mathbb{R}, \tag{6.9}$$

is called ζ-Subbotin distribution.

The family of ζ-Subbotin distributions is closely related to the family of generalized error distributions (GEDs), cf., e.g., Nelson (1991) and references therein. In fact, the Lebesgue density of the GED with shape parameter equal to ζ is a scaled version of the ζ-Subbotin density f_ζ. In case of $\zeta = 2$, both distributions coincide with the standard normal. The 1-Subbotin distribution is equal to the Laplace (or double-exponential) distribution, while the GED with shape parameter equal to 1 has the same shape, but lighter tails.

Theorem 6.2 (Neuvial and Roquain (2012)). *Assume that the (conditional) distribution of X_1 on $\mathbb{R}$, given $Y_1 = 0$, is the ζ-Subbotin distribution with Lebesgue density f_ζ as in* (6.9) *and that the (conditional) distribution of X_1 given $Y_1 = 1$ is a shifted or scaled ζ-Subbotin distribution with Lebesgue density given by $f_{shift}(x) = f_\zeta(x - \mu_m)$ or $f_{scaled}(x) = f_\zeta(x/\sigma_m)/\sigma_m$, where $(\mu_m)_{m\in\mathbb{N}}$ or $(\sigma_m)_{m\in\mathbb{N}}$, respectively, is a sequence of unknown parameters. For all $m \in \mathbb{N}$, assume that μ_m or σ_m, respectively, is such that the density of the (random) p-value p_i corresponding to X_i has under $Y_i = 1$ a continuously decreasing Lebesgue density f_m, fulfilling $f_m(0^+) > \tau_m > f_m(1^-)$, where*

$$\tau_m = \frac{\pi_0(m)}{\pi_1(m)} = m^\beta, \quad 0 < \beta \leq 1.$$

Denoting the cdf corresponding to the p-value density f_m by F_m, assume that there exist constants C_- and C_+ such that $0 < C_- \leq F_m(f_m^{-1}(\tau_m)) \leq C_+ < 1$. Let the FDR level $\alpha = \alpha_m$ in the definition of φ^{LSU} be chosen such that $\alpha_m \to 0$ and $\log(\alpha_m) = o((\log m)^\gamma)$ as $m \to \infty$, where $\gamma = 1 - 1/\zeta$ for $\zeta > 1$ in case of shift alternatives and $\gamma = 1$ for $\zeta \geq 1$ in case of scale alternatives. Then, φ^{LSU} is asymptotically optimal in the sense that it fulfills

$$R_m(\varphi^{LSU}) \sim R_m^{opt}, \quad m \to \infty. \tag{6.10}$$

In (6.10)*, R_m is either one of the risk measures introduced in* (6.3) *and* (6.4)*, and R_m^{opt} is the corresponding risk of the Bayes-optimal classifier with respect to R_m (which thresholds p-values at the fixed cutoff $f_m^{-1}(\tau_m)$).*

In addition, Neuvial and Roquain (2012) derived exact convergence rates at which the relative excess risk $(R_m(\varphi^{LSU}) - R_m^{opt})/R_m^{opt}$ vanishes as $m \to \infty$. As in Theorem 6.1, also under the assumptions of Theorem 6.2 it turns out that φ^{LSU}, regarded as a classifier, is highly adaptive to the amount of sparsity in the data, because the assertion holds true for any $0 < \beta \leq 1$. However, the fine-tuning of the nominal FDR level $\alpha = \alpha_m$ is a bottleneck in practice. Both under the model considered in Theorem 6.1 and under that considered in Theorem 6.2, even the (asymptotically) optimal order of magnitude of $\alpha = \alpha_m$ depends on unknown model parameters.

Remark 6.1.

(a) The risk measures $R^{(T)}(\hat{h}_m)$ and $R^{(I)}(\hat{h}_m)$ from part (b) of Definition 6.1 are defined without a weighting by a cost parameter c. As argued by Neuvial and Roquain (2012), the results of Theorem 6.2 remain to hold true if such a weighting is considered in $R^{(T)}(\hat{h}_m)$ and $R^{(I)}(\hat{h}_m)$.
(b) The sparsity assumptions regarding $\pi_1 = \pi_1(m)$ in Theorems 6.1 and 6.2 are appropriate for signal detection problems, where a small amount of signals (corresponding to $Y_i = 1$) is assumed within a huge amount of data points.
(c) Some further analytical results on FDR-controlled classification can be found in the works by Scott et al. (2009) and Genovese and Wasserman (2004). Cohen and Sackrowitz (2005a, b) studied the classes of single-step, step-down and step-up multiple tests with respect to admissibility and Bayes optimality in the classification context. In particular, they showed that step-up tests like φ^{LSU} are in general inadmissible under additive loss, meaning that uniformly better (and feasible) classification procedures exist, in particular in non-sparse models.

6.2 Binary Classification in Non-Sparse Models

For applications in which the class probabilities π_0 and π_1 are assumed to be (roughly) balanced, as for instance in brain-computer interfacing research that we will consider in Chap. 12, the Bayes risk decomposition given in (6.2) suggests to study multiple tests that control a weighted average of type I and type II error rates. In this direction, Storey (2003) pointed out that among the sets considered in (6.5) there is also a rejection region that minimizes the weighted average of the pFDR and its type II analogue, the positive false non-discovery rate (pFNR). This means, for a given weight parameter $w \in (0, 1)$ it exists a constant $c(w)$ such that

$$\min_{\Gamma \subset \mathbb{R}^k} (A(w)) = (1 - w) \cdot \mathrm{pFDR}(\Gamma_{c(w)}) + w \cdot \mathrm{pFNR}(\Gamma_{c(w)}), \quad \text{where} \tag{6.11}$$

$$A(w) = (1 - w) \cdot \mathrm{pFDR}(\Gamma) + w \cdot \mathrm{pFNR}(\Gamma). \tag{6.12}$$

Under the distributional assumptions of Definition 6.1, $\mathrm{pFDR}(\Gamma)$ and $\mathrm{pFNR}(\Gamma)$ are given by

$$\text{pFDR}(\Gamma) = \mathbb{P}(H_0|X_1 \in \Gamma) = \mathbb{E}\left[\frac{V_m}{R_m}|R_m > 0\right],$$

$$\text{pFNR}(\Gamma) = \mathbb{P}(H_1|X_1 \notin \Gamma) = \mathbb{E}\left[\frac{T_m}{W_m}|W_m > 0\right],$$

where V_m, R_m, T_m and W_m are again as in Table 1.1 and refer to a multiple test with fixed rejection region Γ for every marginal test. A particularly convenient scalability property is that pFDR(Γ) and pFNR(Γ) do not depend on m, in contrast to $\mathbb{E}[V_m]$ and $\mathbb{E}[T_m]$.

In practice, it remains to determine or at least to approximate the optimal cost parameter $c(w)$. Several possible methods for this have been discussed in the literature. As typical in the statistical learning context, many methods rely on utilizing a training sample $((x_i^{\text{train}}, y_i^{\text{train}}))_{1 \le i \le m_{\text{train}}}$ with known labels, where these training data points are assumed to be generated independently of $((X_i, Y_i))_{1 \le i \le m}$ from the distribution $\mathbb{P}$. To this end, it is useful to notice that pFDR(Γ) and pFNR(Γ) can be computed in terms of the densities f_0 and f_1 by

$$\text{pFDR}(\Gamma) = \frac{\pi_0 I_0(\Gamma)}{\pi_0 I_0(\Gamma) + \pi_1 I_1(\Gamma)}, \quad \text{pFNR}(\Gamma) = \frac{\pi_1[1 - I_1(\Gamma)]}{\pi_1[1 - I_1(\Gamma)] + \pi_0[1 - I_0(\Gamma)]} \tag{6.13}$$

with $I_j(\Gamma) = \int_\Gamma f_j(\mathbf{u})\lambda^k(d\mathbf{u})$, $j = 0, 1$. Representation (6.13) shows that the Bayes risk defined in (6.1) can be regarded as a local version of the risk functional $A(w)$ from (6.12). Based on these considerations, in Sect. 7 of Storey (2003) the following algorithm for approximating $c(w)$ is outlined.

Algorithm 6.1

1. *Utilizing training data* $((x_i^{train}, y_i^{train}))_{1 \le i \le m_{train}}$, *estimate the pdfs* f_0 *and* f_1 *by* $\hat{f}_j, j = 0, 1$.
2. *Approximate the sets* Γ_c *for given* $c \in (0, 1)$ *by plugging* $\hat{f}_j$ *into* (6.5) *instead of* f_j, $j = 0, 1$. *The prior probability* π_0 *can either be chosen explicitly or also be estimated from the training data.*
3. *Estimate pFDR(Γ_c) and pFNR(Γ_c) by numerical integration in* (6.13) *with* f_j *replaced by* $\hat{f}_j$, $j = 0, 1$.
4. *Choose* $w \in (0, 1)$ *and minimize the numerical approximation of* $(1 - w) \cdot$ *pFDR*$(\Gamma_c) + w \cdot$ *pFNR*(Γ_c) *with respect to* c.

This approach automatically also delivers an estimate of the optimal rejection region $\Gamma_{c(w)}$, see the second step of the algorithm.

Remark 6.2.

(a) Actually, Storey (2003) describes a slightly different approach, namely, to estimate f_0 from training data drawn from the zero class and to estimate the marginal density $f = \pi_0 f_0 + \pi_1 f_1$ from possibly unlabeled data. This relates the statistical model from Definition 6.1 also to the statistical learning task of semi-supervised novelty detection as in Blanchard et al. (2010).

(b) In contrast to the methods discussed in Sect. 6.1, Algorithm 6.1 is not restricted to feature vector dimensionality $k = 1$. In cases with $k > 1$, estimation methods for multivariate densities are applicable. Excellent textbook references for non-parametric density estimation are Silverman (1986) and Härdle et al. (2004).

Dickhaus et al. (2013) demonstrated that Algorithm 6.1 also works for stationary, but non-trivially auto-correlated feature vectors, at least under weak dependency assumptions. This generalization is important for the classification of multivariate time series data, cf. Chap. 12.

A second plausible approach for approximation of $c(w)$ in (6.11) relies on direct estimation of the likelihood ratio λ, which avoids plug-in of estimated densities. In a series of papers (cf. Sugiyama et al. (2009) and references therein), a group of Japanese researchers developed methods for and discussed applications of such direct estimation of density ratios. In Sects. 2.8 and 4 of Sugiyama et al. (2009), especially the so-called `uLSIF` algorithm is propagated. Hence, the following alternative algorithm has been investigated by Dickhaus et al. (2013), too.

Algorithm 6.2

1. *Utilizing training data* $((x_i^{train}, y_i^{train}))_{1 \leq i \leq m_{train}}$, *estimate the density ratio* λ *by* $\hat{\lambda}$.
2. *Approximate the set* Γ_c *for given* $c \in (0, 1)$ *by plugging* $\hat{\lambda}$ *instead of* λ *into the right-hand side of* (6.5). *The prior probability* π_0 *can either be chosen explicitly or be estimated from the training data.*
3. *Estimate* $pFDR(\Gamma_c)$ *and* $pFNR(\Gamma_c)$ *by calculating the relative frequencies of events* $\{y_i^{train} = 0\}$ *in the training sub-dataset with* $x_i^{train} \in \Gamma_c$ *and* $\{y_i^{train} = 1\}$ *in the training sub-dataset with* $x_i^{train} \notin \Gamma_c$, *respectively.*
4. *Choose* $w \in (0, 1)$ *and minimize the approximation of* $(1 - w) \cdot pFDR(\Gamma_c) + w \cdot pFNR(\Gamma_c)$ *with respect to* c.

The general finding of Dickhaus et al. (2013) was that Algorithm 6.1 seems to be more time-consuming, but that it had slightly better classification performance than Algorithm 6.2, both on computer-simulated and on real multivariate time series data. In the first step of Algorithm 6.1, the authors employed fixed-width kernel density estimators with Gaussian kernels and empirically sphered data, while in the first step of Algorithm 6.2 the proposed `uLSIF` algorithm of Sugiyama et al. (2009) was used.

An interesting direction for future research would be to study the general class of multiple testing based cost functions of the form

$$(1 - w)g_1(\mathbb{P}^{(V,R)}) + wg_2(\mathbb{P}^{(T,W)}),$$

where g_1 and g_2 are given functionals, with respect to binary classification in non-sparse models.

6.3 Feature Selection for Binary Classification via Higher Criticism

In cases where the feature vector dimension k is larger than 1, the explicit determination of the optimal rejection region Γ_c in (6.5) requires multivariate techniques. The presumably most well-known case is that of Fisher discrimination, meaning that the densities f_0 and f_1 are those of multivariate normal distributions on $\mathbb{R}^k$, $k > 1$, with common covariance matrix Σ, but class-specific mean vectors μ_0 and μ_1 (say). In this case, the Bayes-optimal rejection region is given by

$$\Gamma_c = \left\{ x \in \mathbb{R}^k : \left[x - \frac{1}{2}(\mu_1 + \mu_0) \right]^\top \Sigma^{-1}(\mu_1 - \mu_0) \geq \log\left(\frac{\pi_0(1-c)}{\pi_1 c} \right) \right\}. \tag{6.14}$$

This classification rule has a simple structure, because it is linear in the data. Hence, it is easy to apply in practice, provided that the parameters $\mu_j, j = 0, 1$, and Σ are known. In case of unknown parameters, one typically estimates them from a training sample (cf. the first steps in Algorithms 6.1 and 6.2), leading to the so-called linear discriminant analysis (LDA). However, this approach causes severe issues if $k > m_{\text{train}}$, because in such cases the empirical covariance matrix is not invertible. The latter situation often occurs in modern life sciences, where typically a large set of features is at hand. Motivated by this example, Donoho and Jin (2008) were concerned with the problem of feature selection for classification based on multiple testing. Notice that this has close similarities to the problem of model selection that we will treat in Chap. 7.

Theorem 6.3 (Central limit theorem for order statistics). *Let $U_{1:k}, \ldots, U_{k:k}$ denote the order statistics of k stochastically independent, identically UNI$[0,1]$-distributed random variables $U_1, \ldots, U_k$. Let $q \in (0,1)$ be such that $i/k - q = o(k^{-1/2})$ as $k \to \infty$ for some integer-valued sequence $(i = i(k) : k \in \mathbb{N})$. Then, for all $t \in \mathbb{R}$,*

$$\mathbb{P}\left(\sqrt{k} \frac{U_{i:k} - q}{\sqrt{q(1-q)}} \leq t \right) \to \Phi(t), \quad k \to \infty.$$

Proof. See, for instance, Chap. 4 of Reiss (1989). □

Loosely formulated, the assertion of Theorem 6.3 means that for given $1 \leq i \leq k$, where k is large, $U_{i:k}$ is approximately normally distributed with mean i/k and variance $(i/k(1-i/k))/k$. It seems that John Wilder Tukey was the first who suggested to apply this result to multiple test problems with k marginal p-values which are under the global hypothesis H_0 distributed as $U_1, \ldots, U_k$ in Theorem 6.3, see Donoho and Jin (2004) and references therein.

Definition 6.4. (Higher criticism). Let $p_{1:k}, \ldots, p_{k:k}$ denote ordered marginal p-values for a multiple test problem $(\mathcal{X}, \mathcal{F}, \mathcal{P}, \mathcal{H}_k)$. Then, the higher criticism

(HC) objective at index $1 \leq i \leq k$ is given by

$$HC(i, p_{i:k}) = \sqrt{k} \frac{i/k - p_{i:k}}{\sqrt{p_{i:k}(1 - p_{i:k})}}. \tag{6.15}$$

Alternatively and asymptotically equivalently, one may use i/k instead of $p_{i:k}$ in the denominator of $HC(i, p_{i:k})$, see Donoho and Jin (2008). For a given tuning parameter $\lambda \in (0, 1)$, the HC test statistic is given by

$$HC_k^* = \max_{1 \leq i \leq \lambda k} HC(i, p_{i:k}). \tag{6.16}$$

Asymptotic ($k \to \infty$) distributional results concerning HC_k^* have been derived by Donoho and Jin (2004). These results allow for utilizing HC_k^* as a test statistic for the global hypothesis H_0 in $\mathscr{H}_k$, provided that the number k of hypotheses is large. For the specific task of feature selection (where the number k of features is large), Donoho and Jin (2008) proposed the following algorithm.

Algorithm 6.3 *Under the assumptions of Definition 6.1, assume that $k >> 1$ and that a training sample $((x_i^{train}, y_i^{train}))_{1 \leq i \leq m_{train}}$ as described before Algorithm 6.1 is at hand. Furthermore, assume that there are some features (corresponding to components of the vector X_1) which are actually uninformative for the classification task. Then, selection of the informative features can be performed as follows.*

1. *For every feature $1 \leq j \leq k$, construct a statistic $Z_j : \mathbb{R}^{m_{train}} \times \{0, 1\}^{m_{train}} \to \mathbb{R}$ such that $Z = (Z_1, \ldots, Z_k)^\top$ is an (at least asymptotically) Gaussian random vector with stochastically independent components and mean vector $\mu = (\mu_1, \ldots, \mu_k)^\top$, where $\mu_j = 0$ if and only if feature j is uninformative for the classification task.*
2. *For all $1 \leq j \leq k$, compute the p-value p_j corresponding to the two-sided Z-test of the hypothesis $H_j : \{\mu_j = 0\}$ based on Z_j.*
3. *With these p-values, evaluate $HC(j, p_{j:k})$ for all $1 \leq j \leq k$, as well as HC_k^*, see Definition 6.4. Denote the index yielding the maximum in the definition of HC_k^* by j^*.*
4. *Select those features j for which $|Z_j|$ exceeds $|Z_{j^*}|$.*

Under certain assumptions regarding the asymptotic ($k \to \infty$) order of magnitude of the (common) mean of those random variables Z_j for which feature j is informative and the proportion of informative features, Donoho and Jin (2008) demonstrated (and outlined a rigorous proof) that Algorithm 6.3 leads to asymptotically optimal error rate classifiers.

Acknowledgments Parts of Sect. 6.2 originated from joint work with the Berlin brain-computer interface group, in particular with Benjamin Blankertz and Frank C. Meinecke. I thank Gilles Blanchard, Etienne Roquain and Masashi Sugiyama for fruitful discussions.

References

Abramovich F, Benjamini Y, Donoho DL, Johnstone IM (2006) Adapting to unknown sparsity by controlling the false discovery rate. Ann Stat 34(2):584–653, doi:10.1214/009053606000000074.

Berger JO (1985) Statistical decision theory and Bayesian analysis, 2nd edn. Springer-Verlag, Springer Series in Statistics. New York etc.

Blanchard G, Lee G, Scott C (2010) Semi-supervised novelty detection. Journal of Machine Learning Research 11:2973–3009

Bogdan M, Chakrabarti A, Frommlet F, Ghosh JK (2011) Asymptotic Bayes-optimality under sparsity of some multiple testing procedures. Ann Stat 39(3):1551–1579, doi:10.1214/10-AOS869.

Cohen A, Sackrowitz HB (2005a) Characterization of Bayes procedures for multiple endpoint problems and inadmissibility of the step-up procedure. Ann Stat 33(1):145–158, doi:10.1214/009053604000000986.

Cohen A, Sackrowitz HB (2005b) Decision theory results for one-sided multiple comparison procedures. Ann Stat 33(1):126–144, doi:10.1214/009053604000000968.

Dickhaus T, Blankertz B, Meinecke FC (2013) Binary classification with pFDR-pFNR losses. Biom J 55(3):463–477, doi:10.1002/bimj.201200054.

Donoho D, Jin J (2004) Higher criticism for detecting sparse heterogeneous mixtures. Ann Stat 32(3):962–994, doi:10.1214/009053604000000265.

Donoho D, Jin J (2008) Higher criticism thresholding: Optimal feature selection when useful features are rare and weak. Proc Natl Acad Sci USA 105(39):14,790–14,795.

Efron B, Tibshirani R (2002) Empirical Bayes methods and false discovery rates for microarrays. Genet Epidemiol 23(1):70–86

Genovese C, Wasserman L (2004) A stochastic process approach to false discovery control. Ann Stat 32(3):1035–1061, doi:10.1214/009053604000000283.

Härdle W, Müller M, Sperlich S, Werwatz A (2004) Nonparametric and semiparametric models. Springer, Springer Series in Statistics. Berlin

Hastie T, Tibshirani R, Friedman J (2009) The elements of statistical learning. Data mining, inference, and prediction. 2nd ed. Springer Series in Statistics. New York, NY: Springer., doi:10.1007.b94608

Nelson DB (1991) Conditional heteroskedasticity in asset returns: A new approach. Econometrica 59(2):347–370, doi:10.2307/2938260.

Neuvial P, Roquain E (2012) On false discovery rate thresholding for classification under sparsity. Ann Stat 40(5):2572–2600

Reiss RD (1989) Approximate distributions of order statistics. Springer, With applications to non-parametric statistics. Springer Series in Statistics. New York etc.

Scott C, Bellala G, Willett R (2009) The false discovery rate for statistical pattern recognition. Electronic Journal of Statistics 3:651–677

Silverman B (1986) Density estimation for statistics and data analysis. Chapman and Hall, Monographs on Statistics and Applied Probability. London - New York

Storey JD (2003) The positive false discovery rate: A Bayesian interpretation and the q-value. Ann Stat 31(6):2013–2035

Subbotin MT (1923) On the law of frequency of errors. Matematicheskii Sbornik 31(2):296–301

Sugiyama M, Kanamori T, Suzuki T, Hido S, Sese J, Takeuchi I, Wang L (2009) A Density-ratio Framework for Statistical Data Processing. IPSJ Transactions on Computer Vision and Application 1:183–208

Vapnik VN (1998) Statistical learning theory. Adaptive and Learning Systems for Signal Processing, Communications, and Control. Chichester: Wiley.

Chapter 7
Multiple Testing and Model Selection

Abstract This chapter deals with interrelations between multiple testing and model selection. We describe (modifications of) classical multiple test procedures that lead to consistent model selection. Optimization of information criteria is formalized as a multiple test problem and addressed by discussing appropriate multiple testing-based thresholding schemes. Furthermore, the important topic of multiple testing after model selection is treated. We provide distributional results concerning regularized estimators and present two-stage procedures which apply multiple testing in a stage of analysis following the model selection stage. Finally, we are concerned with the problem of selective inference, meaning that model selection and multiple testing is performed in parallel.

Model selection is a highly relevant step in the statistical analysis of high-dimensional data as typically generated by applications from modern life sciences. The problem of model selection occurs if a set of possible models, which are in concurrence to each other, is assumed and the ascertained data shall not (only) be utilized to calibrate these models, but (also) to choose from this set the model that describes the data best according to some pre-specified criteria. In particular, parsimonious models are often preferred, meaning that, in addition to mere model fit characteristics, the complexity of the chosen model shall be as small as possible. This is a challenge in high-dimensional settings where often a large set of potential covariates is considered, but it is assumed that only a small to moderate number of them actually have an influence on the (mean) response. A nice recent overview of model selection techniques for such high-dimensional feature spaces is provided in the invited review article by Fan and Lv (2010). Here, we only mention methods that have a close connection to multiple testing methods that we have discussed in previous chapters, and we discuss the problem of simultaneous inference after or along with model selection (also referred to as selective inference in the literature). In particular, despite their practical relevance, we exclude model selection methods based on cross-validation (cf. Shao (1993, 1997)).

Many of the methods considered in this chapter have originally been derived under the assumption of a linear model in the sense of Definition 4.7, meaning that

T. Dickhaus, *Simultaneous Statistical Inference*,
DOI: 10.1007/978-3-642-45182-9_7,

real-valued, stochastically independent variables $Y = (Y_1, \ldots, Y_n)^\top$ are observed such that

$$Y = X\vartheta + \varepsilon. \tag{7.1}$$

In this, X denotes the $(n \times k)$ design matrix. In contrast to Chap. 4, we consider here the cases of fixed and of random designs. In the latter case, X is a random matrix the columns of which correspond to realizations of predictors (covariates) $X_1, \ldots, X_k$. Also, we consider cases where the dimensionality k may exceed the sample size n. In any case, $\vartheta \in \mathbb{R}^k$ is the target of statistical inference and ε with values in $\mathbb{R}^n$ denotes a vector of random errors or noise variables. A typical distributional assumption is that $\varepsilon \sim \mathcal{N}_n(0, \sigma^2 I_n)$, but for many asymptotic results the assumption of stochastically independent, identically distributed, centered error terms with bounded second moment σ^2 is sufficient.

7.1 Multiple Testing for Model Selection

In a generic manner, Bauer et al. (1988) describe the model selection problem as follows. Under a statistical model $(\mathcal{X}, \mathcal{F}, (\mathbb{P}_\vartheta)_{\vartheta \in \Theta})$ (that may contain further nuisance parameters which we suppress notationally), assume that $\Theta = \mathbb{R}^k$ such that $\vartheta = (\vartheta_1, \ldots, \vartheta_k)^\top$. Typical examples of such parameters are vectors of regression coefficients in a linear model of the form (7.1) or in a generalized linear model, see Definition 4.8. Now, assume that there exists a subset of indices $I_0 \subseteq \{1, \ldots, k\}$ such that $\vartheta_i = 0$ for all $i \in I_0$. Then, the problem of model selection consists in estimating I_0 from the data, i. e., to construct an estimator $\hat{I}_0 : \mathcal{X} \to 2^{\{1,\ldots,k\}}$ for I_0 based on the data $x \in \mathcal{X}$, according to some decision-theoretic criteria. In the linear model context, $\vartheta_i = 0$ has the interpretation that the i-th covariate, which corresponds to column i in the design matrix, has no effect on the (mean) response. Thus, also the term variable selection is often used instead of model selection. Formulating the model selection problem in this way, its close relationship to multiple testing becomes clear. Namely, we consider the system $\mathcal{H} = (H_i : 1 \leq i \leq k)$ of hypotheses, where $H_i : \{\vartheta_i = 0\}$ with alternative $K_i : \{\vartheta_i \neq 0\}$. Then, any multiple test $\varphi = (\varphi_i : 1 \leq i \leq k)$ can also be regarded as an estimator $\hat{I}_0$ by setting $\hat{I}_0(x) = \{1 \leq i \leq k : \varphi_i(x) = 0\}$. Indeed, one class of model selection procedures is targeted towards control of the FWER. In the model selection context, FWER control means to construct $\hat{I}_0$ or φ, respectively, such that

$$\mathbb{P}_\vartheta\left(|\hat{I}_1 \cap I_0| \geq 1\right) \leq \alpha, \quad \hat{I}_1 = \{1, \ldots, k\} \setminus \hat{I}_0, \tag{7.2}$$

where this condition may only be fulfilled asymptotically (for the sample size n tending to infinity); cf., e.g., Wasserman and Roeder (2009) and Meinshausen et al. (2009).

Somewhat contrarily, a model selection procedure is referred to as conservative, if

$$\limsup_{n\to\infty} \mathbb{P}_{\vartheta}\left(\hat{I}_1 \supseteq I_1\right) = 1, \tag{7.3}$$

see, for instance, Sect. 7.2 in Leeb and Pötscher (2009). Conservativity of $\hat{I}_1$ therefore means that, at least with high probability, no non-zero effect (in terms of the coefficients in ϑ) is excluded from the model. In the testing terminology, both criteria (7.2) and (7.3) can be regarded as multiple type I error criteria (where the interpretation of H_i and K_i may be switched in case of (7.3)) such that standard multiple test calibration can be performed for constructing an estimator $\hat{I}_0$ fulfilling these criteria.

However, there exists at least one further popular model selection criterion which balances type I and type II error probabilities.

Definition 7.1. A model selection procedure $\hat{I}_0$ is called consistent, if

$$\lim_{n\to\infty} \mathbb{P}_{\vartheta}\left(\hat{I}_0 = I_0\right) = 1, \tag{7.4}$$

at least on a suitable subset Θ^* of Θ.

Construction of consistent model selection procedures can therefore not straightforwardly be performed by means of standard multiple tests discussed in previous chapters.

Bauer et al. (1988) derived a single-step multiple testing scheme for the model selection problem as follows.

Theorem 7.1 (Bauer et al. (1988)). *Assume that point estimators $\hat{\vartheta}_i$ for the model parameters ϑ_i exist, $1 \leq i \leq k$. Furthermore, let $\sigma_{i,n}$ denote positive real numbers, with corresponding estimators $\hat{\sigma}_{i,n}$, $1 \leq i \leq k$. For all $1 \leq i \leq k$, let $(c_i(n) : n \in \mathbb{N})$ denote a sequence of real numbers such that $c_i(n) \to \infty$, $n \to \infty$. Consider the following estimator of I_0, derived from a single-step multiple test:*

$$\hat{I}_0 = \{1 \leq i \leq k : |\hat{\vartheta}_i|/\hat{\sigma}_{i,n} \leq c_i(n)\}. \tag{7.5}$$

Then, either of the following two sets of conditions is sufficient for $\hat{I}_0$ being a consistent estimator of I_0.

(a) For all $1 \leq i \leq k$, $\mathbb{E}_{\vartheta}[(\hat{\vartheta}_i - \vartheta_i)^2]/\sigma_{i,n}^2$ is bounded, $\hat{\sigma}_{i,n}/\sigma_{i,n}$ tends to one in probability as $n \to \infty$, and $\sigma_{i,n} c_i(n) \to 0$, $n \to \infty$.

(b) For all $1 \leq i \leq k$, $\hat{\sigma}_{i,n} > 0$ almost surely for all $n \in \mathbb{N}$, $(\hat{\vartheta}_i - \vartheta_i)^2/\hat{\sigma}_{i,n}^2$ converges in distribution and $\hat{\sigma}_{i,n} c_i(n) \to 0$ in probability, $n \to \infty$.

Corollary 7.1. *Assume a multiple linear regression model as in* (7.1). *If the multivariate central limit theorem from Theorem 4.4 applies, a consistent variable selection procedure is given by a family of t-tests $(\varphi_i : 1 \leq i \leq k)$ for the parameters ϑ_i, $1 \leq i \leq k$, where the critical values $c_i(n)$ are diverging to infinity such that $c_i(n) n^{-1/2} \to 0$, $n \to \infty$, for all $1 \leq i \leq k$.*

Along similar lines, Bunea et al. (2006) proved consistency of variable selection procedures in linear models based on standard multiple test procedures, where the dimension k of the parameter space is allowed to grow with n.

Theorem 7.2 (Bunea et al. (2006)). *Assume a linear model of the form considered in (7.1), where* $\mathbb{E}\left[|\varepsilon_1|^{4+\delta}\right] < \infty$ *for some* $\delta > 0$ *and the following three additional assumptions hold true.*

(A1) The dimensionality k *fulfills* $k \leq \sqrt{n}/\log(n)$.
(A2) The quantity $\gamma = \max_{1 \leq j \leq k}(X^\top X)^{-1}_{jj}$ *tends to zero,* $n \to \infty$, *with* $\gamma \leq 1/\log(n)$ *for all* $n \in \mathbb{N}$.
(A3) The quantity $r = \max_{1 \leq j \leq n}(X(X^\top X)^{-1}X^\top)_{jj}$ *fulfills* $rk^2 \to 0$, $n \to \infty$.

Let p-values $(p_i : 1 \leq i \leq k)$ *be defined by* $p_i(x) = 2(1 - \Phi(|T_i(x)|))$, *where*

$$T_i(x) = \hat{\vartheta}_i / \sqrt{Var(\hat{\vartheta}_i)}$$

and $\hat{\vartheta}_i$ *denotes the least squares estimator of* ϑ_i, $1 \leq i \leq k$, *cf. Theorem 4.4.*

Then, the Bonferroni t-test given by Example 3.1 as well as the multiple test φ^{BY} *from Theorem 5.5 operating on these p-values lead to a consistent variable selection procedure, provided that the significance level* $\alpha = \alpha_n \in (0, 1)$ *tends to* 0 *for* $n \to \infty$, *such that* $\alpha_n \geq \exp(-n)$ *for all* $n \in \mathbb{N}$ *and* $k\alpha_n/\log(k) \to 0$, $n \to \infty$.

7.2 Multiple Testing and Information Criteria

A general framework for variable selection based on information criteria has been presented by George (2000), see in particular Sect. 7.2 in that article. An information criterion in the context of the linear model (7.1) penalizes the number of explanatory variables (corresponding to columns in the design matrix X) that are included. Hence, model complexity is penalized (subject to a good fit of the data Y), and parsimonious models are preferred. The general form of an information criterion as given by George (2000) assumes that all possible models (corresponding to subsets of $\{1, \ldots, k\}$) are indexed by an index vector γ, which contains the indices of the explanatory variables (covariates) that are included. Then, for a tuning parameter $\lambda > 0$ (occasionally referred to as the penalization intensity), the information criterion can be expressed as

$$RSS_\gamma^{(\lambda)} = RSS_\gamma / \hat{\sigma}^2_{\text{full}} + \lambda|\gamma|, \tag{7.6}$$

where RSS_γ is the sum of squared residuals under the model indexed by γ, $\hat{\sigma}^2_{\text{full}}$ an estimate of the error variance in the full model (based on the entire design matrix X), and $|\gamma|$ the number of elements in γ. Model selection based on information criteria chooses the model for which $RSS_\gamma^{(\lambda)}$ is minimum. Table 7.1, taken from Zuber and

Table 7.1 Information criteria for model selection

Criterion	Reference	Penalization intensity
AIC	Akaike (1974)	$\lambda = 2$
C_p	Mallows (1973)	$\lambda = 2$
BIC	Schwarz (1978)	$\lambda = \log(n)$
RIC	Foster and George (1994)	$\lambda = 2\log(k)$

Strimmer (2011), lists some choices of the penalization intensity λ that have been proposed in the literature.

As noted by George (2000), AIC and C_p are essentially equivalent. They lead to consistent model selection if the dimensionality of the true model increases with n at a suitable rate. The BIC criterion leads to consistent model selection if the latter dimensionality is fixed. More general penalty functions, depending both on the sample size n and on $|\gamma|$, have been discussed by Zheng and Loh (1995).

Assuming random design, Zuber and Strimmer (2011) related minimization of $RSS_{\gamma}^{(\lambda)}$ to multiple testing. They defined correlation-adjusted (marginal) correlation (CAR) scores as

$$\omega = R^{-1/2} R_{XY} \in \mathbb{R}^k,$$

where R denotes the $k \times k$ correlation matrix of the explanatory variables and $R_{XY} \in \mathbb{R}^k$ the vector of marginal correlations of each of the predictors with the response Y. Then, as shown in Sect. 4.8 of Zuber and Strimmer (2011), minimizing $RSS_{\gamma}^{(\lambda)}$ is (at least approximately) equivalent to thresholding the squared empirical CAR scores at the fixed value $\omega_c^2 = \lambda(1 - r^2)/n$, where r^2 denotes the coefficient of determination in the full model. Predictors for which $\hat{\omega}_j^2$ is smaller than ω_c^2 are removed from the full model by the information criterion. In cases with $k > n$, where the empirical correlation matrix $\hat{R}$ is not invertible, the authors propose to replace $\hat{R}$ by a shrinkage estimator, cf., e. g., Schäfer and Strimmer (2005).

By these considerations, (approximate) minimization of $RSS_{\gamma}^{(\lambda)}$ can equivalently be expressed as a single-step multiple test. It is therefore near at hand to investigate further multiple testing schemes for these scores or, again equivalently, to choose the penalty parameter $\lambda = \lambda(k, n, |\gamma|)$ adaptively based on a (stepwise rejective) multiple test procedure. This proposal has been advocated by Abramovich et al. (2006) and Benjamini and Gavrilov (2009). Specifically, Abramovich et al. (2006) suggested the penalty

$$\lambda \equiv \lambda(k, |\gamma|) = \sum_{\ell=1}^{|\gamma|} \Phi^{-2}\left(\frac{\alpha}{2}\frac{\ell}{k}\right), \tag{7.7}$$

where $\Phi^{-2}(\beta)$ denotes the squared upper β-quantile of the standard normal distribution and $\alpha \leq 1/2$ is a given constant. Utilizing this penalty in a backward elimination scheme (i. e., starting with the full model, decide to remove covariates step-by-step, until the first local minimum of $RSS_{\gamma}^{(\lambda)}$ is found) is equivalent to applying the linear step-up test φ^{LSU} from Definition 5.6 at FDR level α to p-values which correspond to the squared standardized estimators $\hat{\vartheta}_\ell^2/\text{Var}(\hat{\vartheta}_\ell)$, $1 \leq \ell \leq k$. For a wide range of true model dimensionalities, Abramovich et al. (2006) showed that this selection rule is asymptotically optimal in the minimax sense. By means of computer simulations, Benjamini and Gavrilov (2009) demonstrated that forward selection of covariates (i.e., starting with the null model, decide to include covariates step-by-step, until the first local minimum of $RSS_\gamma^{(\lambda)}$ is found) in connection with the penalty

$$\lambda \equiv \lambda(k, |\gamma|) = \sum_{\ell=1}^{|\gamma|} \Phi^{-2}\left(\frac{\alpha}{2}\frac{\ell}{k+1-\ell(1-\alpha)}\right) \tag{7.8}$$

performs well over a broad range of parameter settings. This procedure is equivalent to applying the step-down test with AORC-based critical values given by (5.22) with $\beta_m \equiv 1$ at FDR level α to the aforementioned p-values.

7.3 Multiple Testing After Model Selection

In this section, we are considered with the problem of effect size quantification after model selection. More specifically, assume that a model selection procedure $\hat{I}_1$ is applied to the data sample at hand and $\hat{m}_1 = |\hat{I}_1(x)|$ parameters get selected for the observed data x. Then, a problem of practical interest is to assign a p-value to each of the $\hat{m}_1$ selected components or to construct a (simultaneous) confidence region for ϑ based on point estimators $\hat{\vartheta}_i$, $i \in \hat{I}_1(x)$. To this end, at least two different strategies are possible: (a) employ a multivariate (regularized) estimation technique that implicitly performs model selection by estimating some (potentially many) of the ϑ_i, $1 \leq i \leq k$, to be exactly zero, (b) apply first $\hat{I}_1$ to the sample and estimate $(\vartheta_i : i \in \hat{I}_1(x))$ in a second step of analysis.

7.3.1 Distributions of Regularized Estimators

One particularly popular method for penalized regression is the LASSO (least absolute shrinkage and selection operator) introduced by Tibshirani (1996) and motivated as a Bayesian maximum a posteriori estimator by Park and Casella (2008). In contrast to the rest of this work, we denote an observation in a regression analysis, i.e., the vector of response values $(y_i : 1 \leq i \leq n)$, by $y \in \mathcal{Y}$ (instead of $x \in \mathcal{X}$) in this section. Denoting the likelihood function of the (regression) model under $\vartheta \in \mathbb{R}^k$ evaluated at the observed data y by $l(\vartheta, y)$, the LASSO estimator is given by

$$\hat{\vartheta}^{(LASSO)} = \arg\min_{\vartheta \in \mathbb{R}^k} \{-\ln(l(\vartheta, y)) + \lambda \|\vartheta\|_1\} \tag{7.9}$$

for given penalization intensity $\lambda > 0$. The estimator $\hat{\vartheta}^{(LASSO)}$ is said to perform an implicit variable selection, because it involves an L_1-penalty for the vector of regression coefficients, which often leads to sparse solutions in the sense that only few estimated components are non-zero. As mentioned before, an interesting and challenging task is to provide post-hoc effect size quantification for these non-zero components in terms of p-values or confidence regions. Notice that the usual (marginal) confidence regions based on Wald- or t-tests are prone not to keep the nominal coverage probability, both because they do not account for multiplicity and because they are not compatible with the estimation procedure in the sense that they are not based on the distribution of the estimator $\hat{\vartheta}^{(LASSO)}$, but on that of the ordinary MLE (or least squares estimator) $\hat{\vartheta}$.

Asymptotic distributions of regularized estimators have been derived by Knight and Fu (2000) under the assumption of a multiple linear regression model as considered in (7.1). Notice that under this type of model, the term $-\ln(l(\vartheta, y))$ is an isotone transformation of the sum of squares of residuals under ϑ, i. e., $\sum_{i=1}^{n}(y_i - x_i\vartheta)^2$, where x_i denotes the i-th row of the design matrix X. In particular, one of the results of Knight and Fu (2000) regarding $\hat{\vartheta}^{(LASSO)}$ is as follows.

Theorem 7.3 (Theorem 2 of Knight and Fu (2000)). *Under the multiple linear regression model from* (7.1), *assume that the following two regularity assumptions hold true.*

(i) $n^{-1} \max_{1 \le i \le n} x_i x_i^\top \to 0$, $n \to \infty$, *where* x_i *denotes the i-th row of the design matrix* X_n *for sample size n.*
(ii) $n^{-1} X_n^\top X_n \to V$, $V \in \mathbb{R}^{k \times k}$ *symmetric and positive definite.*

Then, if the penalization intensity $\lambda = \lambda_n$ *in* (7.9) *fulfills* $\lambda_n/\sqrt{n} \to \lambda_0 \ge 0$, $n \to \infty$, *it holds*

$$\sqrt{n}\left(\hat{\vartheta}^{(LASSO)}(n) - \vartheta\right) \xrightarrow{d} \arg\min(f), \quad n \to \infty, \tag{7.10}$$

where

$$f(u) = -2u^\top W + u^\top V u + \lambda_0 \sum_{j=1}^{k} \left[u_j \operatorname{sgn}(\vartheta_j)\mathbf{1}(\vartheta_j \neq 0) + |u_j|\mathbf{1}(\vartheta_j = 0)\right], \, u \in \mathbb{R}^k,$$

and $W \sim \mathcal{N}_k(0, \sigma^2 V)$.

As argued by the authors, the limiting distribution in (7.10) puts positive point mass at the value $u_j = 0$ if $\vartheta_j = 0$, an interesting property in view of consistent model selection. Conditions for actual model selection consistency of $\hat{\vartheta}^{(LASSO)}$ have been provided by Zhao and Yu (2006).

Pötscher and Leeb (2009) extended the findings of Knight and Fu (2000) by providing the finite-sample distribution of $\hat{\vartheta}^{(LASSO)}$. From the practical point of view, it is inconvenient that the limiting distribution given in Theorem 7.3 itself depends on ϑ. Hence, Knight and Fu (2000) considered bootstrap methods for approximating it. However, a further point worth mentioning is that the assertion of Theorem 7.3

is a pointwise one with respect to $\vartheta \in \Theta$. As shown by Leeb and Pötscher (2006, 2008) and Pötscher and Leeb (2009), uniformly (over Θ) consistent estimators for the cdf of a post-model selection estimator, including $\hat{\vartheta}^{(LASSO)}$, do not exist in general. This is one of the reasons why one may restrict attention to parameter subspaces in Definition 7.1.

Feasible LASSO-based confidence regions for ϑ have been developed by Zhang and Zhang (2014) and Bühlmann (2013) in the context of linear models with Gaussian noise. The general idea is to correct the bias of $\hat{\vartheta}^{(LASSO)}$ (or a related regularized estimator like the solution of ridge regression, where an L_2-penalty is involved). van de Geer et al. (2013) generalized this method to non-Gaussian settings and generalized linear models. For the particular task of multiple testing, Bühlmann (2013) defines asymptotically valid p-values based on bias-corrected ridge regression.

Theorem 7.4 (Bühlmann (2013)). *Assume a multiple linear regression model as in* (7.1)*, where* $\varepsilon \sim \mathcal{N}_n(0, \sigma^2 I_n)$. *The ridge estimator is given by*

$$\hat{\vartheta}^{Ridge} = \arg\min_{\vartheta \in \mathbb{R}^k} \{\|Y - X\vartheta\|_2^2/n + \lambda\, \|\vartheta\|_2^2\}. \tag{7.11}$$

Assuming $k > n$, *define the projection matrix* $P_X = X^\top (XX^\top)^+ X$, *where* $(XX^\top)^+$ *denotes the Moore-Penrose pseudo inverse of* $XX^\top$, *and the bias-corrected ridge estimator by*

$$\hat{\vartheta}_j^{corr.} = \hat{\vartheta}_j^{Ridge} - \sum_{\ell \neq j} (P_X)_{j,\ell} \hat{\vartheta}_\ell^{(LASSO)}, \quad 1 \leq j \leq k. \tag{7.12}$$

Then, a stochastic representation of $\hat{\vartheta}^{corr.}$ *is given by*

$$\frac{\hat{\vartheta}_j^{corr.}}{(P_X)_{j,j}} - \vartheta_j \stackrel{d}{=} \frac{Z_j}{(P_X)_{j,j}} - \sum_{\ell \neq j} \frac{(P_X)_{j,\ell}}{(P_X)_{j,j}} \left(\hat{\vartheta}_\ell^{(LASSO)} - \vartheta_\ell\right) + \frac{b_j(\lambda)}{(P_X)_{j,j}},$$

where $Z = (Z_1, \ldots, Z_k)^\top \sim \mathcal{N}_k(0, \sigma^2 \Omega)$, Ω *denotes the covariance matrix of* $\hat{\vartheta}^{Ridge}$, *and* $b_j(\lambda) = \mathbb{E}_\vartheta[\hat{\vartheta}^{Ridge}] - (P_X\vartheta)_j$. *Consequently, for appropriately chosen penalization intensity* $\lambda = \lambda_n$, *it holds*

$$\forall u \in \mathbb{R} : \limsup_{n \to \infty} \left(\mathbb{P}_\vartheta(a_{n,k;j}(\hat{\sigma})|\hat{\vartheta}_j^{corr.}| > u) - \mathbb{P}_\vartheta(|W| + \Delta_j > u)\right) \leq 0, \tag{7.13}$$

where $a_{n,k;j}(\sigma)^{-1}$ *is the standard deviation of* Z_j, $W \sim \mathcal{N}(0, 1)$, *and* Δ_j *denotes an upper bound such that*

$$\mathbb{P}_\vartheta\left(\bigcap_{j=1}^k \left\{ |a_{n,k;j}(\sigma) \sum_{\ell \neq j} (P_X)_{j,\ell} \left(\hat{\vartheta}_\ell^{(LASSO)} - \vartheta_\ell\right)| \leq \Delta_j \right\}\right) \to 1, \quad n \to \infty.$$

Hence, an asymptotically valid p-value for the hypothesis $H_j : \{\vartheta_j = 0\}$ *with two-sided alternative* $K_j : \{\vartheta_j \neq 0\}$ *based on* (7.13) *is given by*

$$p_j = 2(1 - \Phi((a_{n,k;j}(\hat{\sigma})|\hat{\vartheta}_j^{corr.}| - \Delta_j)_+)).$$

Since $(\Delta_j : 1 \leq j \leq k)$ can be constructed explicitly, Theorem 7.4 provides a convenient way to perform multiple testing or to construct confidence regions after L_2-penalized least squares regression. The methods by Zhang and Zhang (2014) and van de Geer et al. (2013) work in the same spirit.

7.3.2 Two-Stage Procedures

Two-stage procedures that calculate p-values in a data analysis step succeeding a first selection step of promising candidate variables have been discussed by Wasserman and Roeder (2009) and Meinshausen et al. (2009). The drawback of such methods is that they rely on sample splitting such that selection and p-value calculation is performed on independent data sub-samples. While this ensures mathematical validity of such procedures in a straightforward manner, it leads to low power for detecting true effects in practice, due to the drastically reduced sample size for effect size estimation. A promising alternative is given by sample-splitting and rejoining, meaning that the full sample is used to screen candidate variables and the full samples for these screened variables are also used in the second (p-value calculation and multiple testing) step. However, calibration of such sample-splitting and rejoining tests for multiple type I error control of given type and level is extremely challenging, in particular because the subset pivotality condition (see Definition 4.3) is prone to be violated such that resampling under the global hypothesis H_0 may lead to invalid p-value thresholds. In practice, one may conduct extensive computer simulations to assess if a given resampling scheme leads to a conservative p-value threshold.

Berk et al. (2013) showed that the full sample may be used in the second step of analysis, provided that p-values are appropriately adjusted for multiplicity. In this respect, simultaneous inference implies valid post-selection inference. However, notice that, in contrast to Sect. 7.3.1, the underlying interpretation here is that "the coefficients of excluded predictors are not zero; they are not defined and therefore do not exist" (Berk et al. (2013), Sect. 2.1).

Theorem 7.5 (Berk et al. (2013)). *Assume a normally distributed response vector* $Y = (Y_1, \ldots, Y_n)^\top$ *with values in* $\mathbb{R}^n$, *where* $Y \sim \mathcal{N}_n(\mu, \sigma^2 I_n)$. *Assume that the mean vector* μ *is of interest, while* σ^2 *is a nuisance parameter. Let a design matrix* $X \in \mathbb{R}^{n \times k}$ *be given, the columns of which are denoted by* $X_1, \ldots, X_k$. *Any model* $M \in \mathcal{M}$ *is identified with a subset of* $\{1, \ldots, k\}$, *where* $\mathcal{M}$ *is the space of all considered models (not necessarily equal to the power set of* $\{1, \ldots, k\}$*). Denoting* $d = \min\{n, k\}$, *it is assumed that for any model* $M \in \mathcal{M}$, $M = \{j_1, \ldots, j_m\}$, *the submatrix* $X_M = (X_{j_1}, \ldots, X_{j_m})$ *has rank* $m \leq d$. *In model* M, *let* $\vartheta_M = \arg\min_{\beta \in \mathbb{R}^m} \|\mu - X_M\beta\|_2^2$ *denote the target of statistical inference and* $\hat{\vartheta}_M = (X_M^\top X_M)^{-1} X_M^\top Y$ *the least squares estimator of* ϑ_M. *Assume that an estimator* $\hat{\sigma}^2$ *exists which is stochastically independent of all* $\hat{\vartheta}_M$ *and fulfills* $r\hat{\sigma}^2/\sigma^2 \sim \chi_r^2$ *for* r *degrees of freedom. Let the t-ratio for component* j *in model* M *be given by*

$$t_{j\cdot M} = \frac{\hat{\vartheta}_{j\cdot M} - \vartheta_{j\cdot M}}{\left((X_M^\top X_M)^{-1}\right)_{jj}^{1/2} \hat{\sigma}}$$

and define $K = K(X, \mathcal{M}, \alpha, r)$ *as the minimal value that satisfies*

$$\mathbb{P}\left(\max_{M \in \mathcal{M}} \max_{j \in M} |t_{j\cdot M}| \leq K\right) \geq 1 - \alpha, \tag{7.14}$$

where the probability measure $\mathbb{P}$ *in* (7.14) *is pivotal. Then, for any model selection procedure, defined as a measurable map* $\hat{M} : \mathbb{R}^n \to \mathcal{M}$, $Y \mapsto \hat{M}(Y)$, *the following assertions hold true.*

(a) $\mathbb{P}\left(\max_{j \in \hat{M}} |t_{j\cdot \hat{M}}| \leq K\right) \geq 1 - \alpha$.

(b) *Letting* $CI_{j\cdot M}(K) = \left[\hat{\vartheta}_{j\cdot M} \mp K \left((X_M^\top X_M)^{-1}\right)_{jj}^{1/2} \hat{\sigma}\right]$, *it holds for all* μ *and* σ^2 *that* $\mathbb{P}_{(\mu,\sigma^2)}\left(\forall j \in \hat{M} : CI_{j\cdot M}(K) \ni \vartheta_{j\cdot M}\right) \geq 1 - \alpha$.

(c) *For all* μ *and* σ^2, $\mathbb{P}_{(\mu,\sigma^2)}\left(\exists j \in \hat{M} : \vartheta_{j\cdot M} = 0 \text{ and } |t_{j\cdot M}^{(0)}| > K\right) \leq \alpha$, *where* $t_{j\cdot M}^{(0)}$ *denotes the t-statistic for testing* $H_j : \{\vartheta_{j\cdot M} = 0\}$.

Moreover, $K(X, \mathcal{M}, \alpha, r) \leq \sqrt{dF_{d,r;\alpha}}$ *for all* X *and* $\mathcal{M}$. *Hence, a Scheffé correction (cf. Theorem 3.1) ensures valid post-selection inference.*

It appears that the general reasoning of Theorem 7.5 can be applied to a variety of further models, too.

7.4 Selective Inference

Here, we discuss procedures which combine selection and multiple testing. In particular, we assume a model with m real-valued parameters and that the researcher is interested in confidence regions for $k < m$ selected parameters, which typically correspond to the k empirically largest effect sizes estimates. It is clear that the reduced dimensionality should allow for a relaxed multiplicity correction compared with the Bonferroni adjustment α/m, where $1 - \alpha$ is the nominal coverage probability for the confidence region in $\mathbb{R}^k$. However, a Bonferroni-type adjustment of the form α/k will typically not lead to a correct coverage probability, because it ignores the fact that selection has taken place. Along these lines, Qiu and Hwang (2007) proposed simultaneous confidence intervals for selected parameters in Gaussian random effects models.

Theorem 7.6 (Qiu and Hwang (2007)). *Consider the model $X_i = \vartheta_i + \varepsilon_i$, $1 \leq i \leq m$, where $\vartheta_i \sim \mathcal{N}(0, \tau^2)$, $\varepsilon_i \sim \mathcal{N}(0, \sigma^2)$, and ϑ_i and ε_i are stochastically independent. Let $X_{1:m} \leq X_{2:m} \leq \cdots \leq X_{m:m}$ denote the order statistics of the X_i. Assuming σ^2 as known, define for any $1 \leq j \leq m$ and fixed $\alpha \in (0, 1)$ the constant q_j by*

$$\mathbb{P}(|Z| < q_j) = 1 - \alpha/j, \quad Z \sim \mathcal{N}(0, 1)$$

and let

$$CI_{i;k} = \left[\hat{M}X_{i:m} \mp v_k(\hat{M})\right], \text{ where}$$

$$\hat{M} = \max\left(0, 1 - \frac{(N-2)\sigma^2}{\sqrt{\sum_{i=1}^m X_i^2}}\right),$$

$$v_k^2(\hat{M}) = \sigma^2 \hat{M}(q_k^2 - \log(\hat{M}))\mathbf{1}(\hat{M} > 0).$$

Then it holds for any given (non-random) subset $S \subset \{1, \ldots, m\}$ with $|S| = k$ that

$$\forall \tau^2 > 0 : \lim_{m \to \infty} \mathbb{P}_{\tau^2}\left(\forall i \in S : \vartheta_{i:m} \in CI_{i;k}\right) > 1 - \alpha,$$

where $\vartheta_{i:m}$ corresponds to observation $X_{i:m}$.

The case of unknown error variance σ^2 in Theorem 7.6 can be treated by Studentization. For finite m, Qiu and Hwang (2007) proposed to use a truncated version of $\hat{M}$, namely, $\hat{M}_k^* = \max(\hat{M}, M_k)$, where $M_k = 1 - F^{-1}_{\chi^2_m}(\alpha/k)/(m-2)$, ensuring that $\hat{M}_k^* > 0$ with high probability. They also treat the case that ϑ_i has a mixture distribution with point mass in 0 and the case of data-dependent selection rules. By evaluating real data from a microarray experiment, they demonstrate the usefulness of their methodology for gene expression analyses that we will treat in Chap. 10 of the present work. Recently, Hwang and Zhao (2013) extended the methodology of Qiu and Hwang (2007) to cases where error variances may be unequal.

Notice that the methodology of Qiu and Hwang (2007) is targeted towards an analogue of FWER control, where the family is constituted by the selected parameters. Hence, it may be natural to consider also a criterion which is analogous to FDR control, in particular for cases with high-dimensional parameter spaces. The latter approach has been introduced by Benjamini and Yekutieli (2005) (see also Benjamini et al. (2009)).

Definition 7.2. (Benjamini and Yekutieli (2005)). Assume a statistical model with m real-valued parameters ϑ_i, $1 \leq i \leq m$ and corresponding point estimators $T = (T_1, \ldots, T_m)^\top$. Consider any measurable selection rule $\mathcal{S} : \mathbb{R}^m \to 2^{\{1,\ldots,m\}}$, $T \mapsto \mathcal{S}(T) \subseteq \{1, \ldots, m\}$. Denote $R_{CI} = |\mathcal{S}(T)|$ and assume that for a given realization of T exactly the R_{CI} confidence intervals for the ϑ_i with $i \in \mathcal{S}(T)$ are to be constructed. Let $V_{CI} \leq R_{CI}$ denote the (random) number of these R_{CI} confidence intervals which fail to cover the associated parameter value. Then, the false coverage-statement rate

(FCR) is defined by

$$FCR = \mathbb{E}_T \left[\frac{V_{CI}}{\max(R_{CI}, 0)} \right].$$

Theorem 7.7 (Benjamini and Yekutieli (2005)). *Under the assumptions of Definition 7.2, assume that the components of T are jointly stochastically independent. Denote by $p_1, \ldots, p_m$ valid p-values corresponding to $T_1, \ldots, T_m$. Then, the following algorithm controls the FCR at level $\alpha \in (0, 1)$.*

1. *Let $R = \max\{1 \leq i \leq m : p_{i:m} \leq i\alpha/m\}$.*
2. *Select the R parameters for which $p_{i:m} \leq R\alpha/m$.*
3. *For each selected parameter, construct a $(1 - R\alpha/m)$ confidence interval.*

Notice that the selection criterion in Theorem 7.7 means to apply the linear step-up test φ^{LSU} from Definition 5.6 to $p_1, \ldots, p_m$ and to let R denote the observed number of rejections of φ^{LSU}. In this sense, φ^{LSU} is a selection rule which is compatible with FCR control for the selected parameters. Benjamini and Yekutieli (2005) also provided additional conditions under which the algorithm from Theorem 7.7 controls the FCR under the PRDS assumption on T. General dependencies can be addressed by an adjustment factor in analogy to Theorem 5.5. Bayesian and empirical Bayesian approaches to FCR control have been discussed by Yekutieli (2012) and Zhao and Hwang (2012).

References

Abramovich F, Benjamini Y, Donoho DL, Johnstone IM (2006) Adapting to unknown sparsity by controlling the false discovery rate. Ann Stat 34(2):584–653. doi:10.1214/009053606000000074

Akaike H (1974) A new look at the statistical model identification. IEEE Trans Autom Control 19:716–723. doi:10.1109/TAC.1974.1100705

Bauer P, Pötscher BM, Hackl P (1988) Model selection by multiple test procedures. Statistics 19(1):39–44. doi:10.1080/02331888808802068

Benjamini Y, Gavrilov Y (2009) A simple forward selection procedure based on false discovery rate control. Ann Appl Stat 3(1):179–198. doi:10.1214/08-AOAS194

Benjamini Y, Yekutieli D (2005) False discovery rate-adjusted multiple confidence intervals for selected parameters. J Am Stat Assoc 100(469):71–81. doi:10.1198/016214504000001907

Benjamini Y, Heller R, Yekutieli D (2009) Selective inference in complex research. Philos Trans R Soc Lond, Ser A, Math Phys Eng Sci 367(1906):4255–4271 doi:10.1098/rsta.2009.0127

Berk R, Brown L, Buja A, Zhang K, Zhao L (2013) Valid post-selection inference. Ann Stat 41(2):802–837

Bühlmann P (2013) Statistical significance in high-dimensional linear models. Bernoulli 19(4): 1212–1242

Bunea F, Wegkamp MH, Auguste A (2006) Consistent variable selection in high dimensional regression via multiple testing. J Stat Plann Inference 136(12):4349–4364. doi:10.1016/j.jspi.2005.03.011

Fan J, Lv J (2010) A selective overview of variable selection in high dimensional feature space. Stat Sin 20(1):101–148

Foster DP, George EI (1994) The risk inflation criterion for multiple regression. Ann Stat 22(4):1947–1975. doi:10.1214/aos/1176325766

van de Geer S, Bühlmann P, Ritov Y (2013) On asymptotically optimal confidence regions and tests for high-dimensional models. arXiv:1303.0518v1

George EI (2000) The variable selection problem. J Am Stat Assoc 95(452):1304–1308. doi:10.2307/2669776

Hwang JTG, Zhao Z (2013) Empirical Bayes Confidence Intervals for Selected Parameters in High-dimensional Data. J Am Stat Assoc forthcoming

Knight K, Fu W (2000) Asymptotics for Lasso-type estimators. Ann Stat 28(5):1356–1378. doi:10.1214/aos/1015957397

Leeb H, Pötscher BM (2006) Can one estimate the conditional distribution of post-model-selection estimators? Ann Stat 34(5):2554–2591. doi:10.1214/009053606000000821

Leeb H, Pötscher BM (2008) Can one estimate the unconditional distribution of post-model-selection estimators? Econom Theory 24(2):338–376. doi:10.1017/S0266466608080158

Leeb H, Pötscher BM (2009) Model selection. In: Andersen, Torben G et al (eds) Handbook of financial time series. With a foreword by Robert Engle. Springer, Berlin pp 889–925. doi:10.1007/978-3-540-71297-8_39

Mallows C (1973) Some comments on C_p. Technometrics 15:661–675 doi:10.2307/1267380

Meinshausen N, Meier L, Bühlmann P (2009) p-values for high-dimensional regression. J Am Stat Assoc 104(488):1671–1681. doi:10.1198/jasa.2009.tm08647

Park T, Casella G (2008) The Bayesian lasso. J Am Stat Assoc 103(482):681–686. doi:10.1198/016214508000000337

Pötscher BM, Leeb H (2009) On the distribution of penalized maximum likelihood estimators: the LASSO, SCAD, and thresholding. J Multivariate Anal 100(9):2065–2082. doi:10.1016/j.jmva.2009.06.010

Qiu J, Hwang J (2007) Sharp simultaneous confidence intervals for the means of selected populations with application to microarray data analysis. Biometrics 63(3):767–766. doi:10.1111/j.1541-0420.2007.00770.x

Schäfer J, Strimmer K (2005) A shrinkage approach to large-scale covariance matrix estimation and implications for functional genomics. Stat Appl Genet Mol Biol 4: Article 32

Schwarz G (1978) Estimating the dimension of a model. Ann Stat 6:461–464. doi:10.1214/aos/1176344136

Shao J (1993) Linear model selection by cross-validation. J Am Stat Assoc 88(422):486–494. doi:10.2307/2290328

Shao J (1997) An asymptotic theory for linear model selection. (With discussion). Stat Sin 7(2):221–264

Tibshirani R (1996) Regression shrinkage and selection via the lasso. J R Stat Soc, Ser B 58(1):267–288

Wasserman L, Roeder K (2009) High-dimensional variable selection. Ann Stat 37(5A):2178–2201

Yekutieli D (2012) Adjusted Bayesian inference for selected parameters. J R Stat Soc Ser B Stat Methodol 74(3):515–541

Zhang CH, Zhang SS (2014) Confidence intervals for low-dimensional parameters in high-dimensional linear models. J R Stat Soc, Ser B, Stat Methodol forthcoming

Zhao P, Yu B (2006) On model selection consistency of Lasso. J Mach Learn Res 7:2541–2563

Zhao Z, Hwang JTG (2012) Empirical Bayes false coverage rate controlling confidence intervals. J R Stat Soc Ser B Stat Methodol 74(5):871–891

Zheng X, Loh WY (1995) Consistent variable selection in linear models. J Am Stat Assoc 90(429):151–156 doi:10.2307/2291138

Zuber V, Strimmer K (2011) High-dimensional regression and variable selection using CAR scores. Stat Appl Genet Mol Biol 10(1):Article 34

Chapter 8
Software Solutions for Multiple Hypotheses Testing

Abstract As a link between the theoretical Part I and the application-oriented Parts II and III of the present book, this chapter is concerned with software solutions for multiple hypotheses testing. In this, we restrict our attention to packages and routines for the R software environment for statistical computing. In particular, we introduce the R packages multcomp and multtest, as well as the R-based μTOSS software. The μTOSS software provides unifying interfaces to multcomp and multtest functions, as well as a graphical user interface. Virtually all multiple tests that are theoretically described in Part I of this book are implemented in the aforementioned software packages. Hence, real-life datasets in Parts II and III can and will be analyzed by employing μTOSS.

For a smooth transition to Parts II and III of our work, we review some software solutions for multiple hypotheses testing in this chapter. In this, we restrict our attention to routines and packages for the software environment R for statistical computing. The R software has become a quasi-standard in research and education, because it is source-open and freely available for many operating systems. Furthermore, many contributors from all fields of statistics regularly extend the R software by writing and publishing individual R packages which implement specific statistical procedures. A quality control for these contributed packages is provided by the Comprehensive R Archive Network (CRAN). Before a new contributed package is released on the CRAN server, it has to pass a variety of test runs.

The real data applications that we are going to discuss in the following chapters of the present work will be evaluated by making use of software that we describe here. In particular, virtually all multiple tests that we have theoretically described in previous chapters are implemented in the following software packages.

T. Dickhaus, *Simultaneous Statistical Inference*,
DOI: 10.1007/978-3-642-45182-9_8,

8.1 The R Package multcomp

The multcomp package provides FWER-controlling simultaneous test procedures for testing linear hypotheses in (rather) general parametric models which fulfill certain regularity assumptions entailing a central limit theorem for an estimator $\hat{\vartheta}$ of the parameter ϑ of interest, cf., among many others, Hothorn et al (2008) and our Sect. 4.2. The backbone of all multcomp procedures is constituted by numerical integration routines for multivariate t- and normal distributions taken from the mvtnorm package, see the book by Genz and Bretz (2009) for details. A multcomp manual with several real data examples, together with some methodological descriptions of the covered multiple test procedures, has been published by Bretz et al. (2010).

The general mechanism for computing a multiplicity-adjusted critical value c_α (say) to which the components of an estimator or a (linear) transformation thereof have to be compared is in the multcomp package (in a nutshell) as follows: a broad class of model objects in R allows for estimating the vector ϑ of model parameters by some estimator $\hat{\vartheta}$, for instance by maximum likelihood or by the method of moments. The estimated coefficients $\hat{\vartheta}_1, \ldots, \hat{\vartheta}_k$ can be read out by utilizing the coef function. In addition, a numerical approximation of the (limiting) covariance matrix of the estimator $\hat{\vartheta}$ is available by utilizing the vcov function. Then, assuming that the model fulfills the regularity assumptions entailing a central limit theorem for $\hat{\vartheta}$, the results of these two function calls are piped into the appropriate function from the mvtnorm package which in turn computes c_α in a numerically stable and efficient manner. The linear hypotheses that have to be tested can be specified by means of contrast matrices and right-hand side vectors as described around (4.7). Some standard contrast matrices (corresponding, for example, to the problems of multiple comparisons with a control or all pairwise comparisons) are already predefined and available in the software. For this entire workflow, the rather generic wrapper routine glht (general linear hypotheses testing) is convenient to use.

A further convenient feature of the multcomp package is the possibility to illustrate the decision pattern of the multiple test procedure graphically by the compact letter display based on the algorithm by Piepho (2004). This display facilitates the interpretation and the communication of the test results. The corresponding multcomp routine for plotting the compact letter display is called plot.cld.

8.2 The R Package multtest

The package multtest is a valuable tool for performing resampling-based multiple hypotheses testing according to the methods described by Westfall and Young (1993) and Dudoit and van der Laan (2008), respectively. Since the package is no longer available via CRAN, but has been integrated into the Bioconductor bundle, it has to be installed via the R console by typing the following commands.

```
source("http://bioconductor.org/biocLite.R")
biocLite("multtest")
```

The three main functions in the `multtest` package are `mt.maxT` and `mt.minP` on the one hand and `MTP` on the other hand. The functions `mt.maxT` and `mt.minP` implement the maxT and minP permutation procedures by Westfall and Young (1993), respectively. With respect to the marginal models per individual comparison, two-sample t-tests (for equal variances and according to Welch (1938)), the nonparametric Mann-Whitney-Wilcoxon two-sample test (cf. Mann and Whitney (1974), Wilcoxon (1945)) and different F-tests for cases with more than two groups are available, together with nonparametric counterparts like the Kruskal-Wallis test and the Friedman test (see, for instance, Chaps. 6 and 7 in the textbook by Hollander and Wolfe (1973) for details of the latter tests).

The function `MTP` can be used for a variety of multiple one- and k-sample tests that have been described by Dudoit and van der Laan (2008) and in the references therein. In addition to permutation-based methods, the latter function also includes several nonparametric bootstrap procedures. Furthermore, the `multtest` package also implements a variety of p-value based multiple tests that we have described in Chaps. 3 and 5, including Bonferroni and Šidák corrections (cf. Examples 3.1 and 3.2), Bonferroni-Holm and Šidák-Holm tests (see Definition 5.3) and Hommel's step-up test (see Definition 5.4) for FWER control, as well as linear step-up tests for control of the false discovery rate as described in Definition 5.6 and Theorem 5.5, among others. The way of implementation of these p-value based procedures in the `multtest` package is by means of adjusted p-values in the function `mt.rawp2adjp`. This means that the function `mt.rawp2adjp` takes marginal, "raw" (unadjusted) p-values $p_1, \ldots, p_m$ as input and transforms them into adjusted p-values, such that the adjusted p-value $p_i^{\text{adj.}}$ (say) corresponding to the original p-value p_i, where $1 \leq i \leq m$, is smaller than α if and only if the chosen multiple test procedure would reject hypothesis H_i at FWER- or FDR-level α, respectively.

8.3 The R-based μTOSS Software

The μTOSS (multiple hypothesis testing in an open software system) software has been programmed in 2010 at Berlin Institute of Technology as a project within the Harvest programme of the PASCAL2 European Network of Excellence. It is constituted by the two R packages `mutoss` (for running multiple tests on the R console) and `mutossGUI` (implementing a graphical user interface, see Sect. 8.3.2 below for details). This provides an easy-to-extend platform for multiple hypotheses testing.

For researchers, the console oriented `mutoss` package features a convenient unification of interfaces for multiple test procedures, in particular including standardized wrapper functions to access the routines from the `multcomp` and `multtest` packages that we have described in Sects. 8.1 and 8.2, respectively. Furthermore, the software provides implementations of recent multiple test procedures that are to the best of our knowledge not available elsewhere, for example AORC-based step-up-down tests, see Sect. 5.5. Every included method comes with a precise description of its usage, its assumptions and appropriate references, cf. our discussion around Fig. 8.3 below. This is meant to be of help both to programmers who want to extend μTOSS and to end users who are typically not experts in the vast diversity of existing multiple test procedures.

Another convenient feature of μTOSS is that it provides helper functions facilitating the setup of benchmark simulations for comparison of competing methods, see Sect. 8.3.1. An extended user's guide to the μTOSS software has been published by Blanchard et al. (2010). In particular, Blanchard et al. (2010) explain in detail how exactly to use the aforementioned wrapper functions for accessing existing `R` packages for multiple testing via μTOSS and how to integrate own `R` implementations of specific multiple tests into the μTOSS software and into the graphical user interface automatically, i. e., without having to contact the μTOSS programmers.

8.3.1 The μTOSS Simulation Tool

It is fair to say that up to now every research group in the multiple testing community uses their own implementations, making (simulation) study evaluations and related results not entirely comparable. Hence, a standardization of input and output parameters for multiple test procedures of the same type (for instance, step-up-down tests) is of primary importance in order to set up comparison benchmarks between different procedures in an easy manner. This is of use both for users wanting to explore the output of different methods on a given dataset, and for developers of new methodology who want to compare the performance of their method against reference procedures on simulated data.

Given the importance of the latter use cases, functions for facilitating the setup of large computer simulations are included as part of μTOSS. The simulation platform consists of just two functions, `simulation()` and `gatherStatistics()`, which are essentially automating loop work for the user.

To illustrate the simplicity of this approach, consider the following `R` code.

```
##############################
#    Perform simulations     #
##############################
my_sim <- simulation(replications = 1000,
            DataGen = list(funName="AR1",
            fun=generate_AR1_p_values,
            m=my_m, m0=my_m0, rho=my_rho),
            listOfProcedures =
            list(list(funName="Hommel",
                      fun = hommel,
```

```
                          alpha = 0.05,
                          silent = TRUE),
                     list(funName="LSU",
                     fun=BH, alpha = 0.05,
                     silent = TRUE)))

result <- gatherStatistics(my_sim,
              listOfStatisticFunctions =
              list(V.ge.0 = V.greater.Zero, FDP=FDP),
              listOfAvgFunctions = list(MEAN = mean))


#############################
#     Simulation results    #
#############################
print(result)
```

Without going into too much detail, the aim of this computer simulation is to compare the performance of Hommel's step-up test (see Definition 5.4) for FWER control and the linear step-up test for control of the false discovery rate (see Definition 5.6) in the case that marginal p-values $p_1, \ldots, p_m$ are generated by some autoregressive time series model of order 1, the implementation of which is not displayed here, but hidden in the function `generate_AR1_p_values`. In this, the number m_0 of true hypotheses and the value ρ of the AR(1) parameter vary, whereas the total number $m = 5$ of hypotheses to be tested stays fixed. After the execution of the two `mutoss` functions `simulation()` and `gatherStatistics()` with the respective parameter lists as input as shown in the R code from above, the `mutoss` software automatically outputs a list of the simulation results as follows (only partly shown).

```
$statisticDF
   funName m m0  rho method alpha  V.ge.0.MEAN   FDP.MEAN
1      AR1 5  1  0.1 Hommel  0.05        0.041 0.00820000
2      AR1 5  1  0.1    LSU  0.05        0.041 0.00820000
3      AR1 5  3  0.1 Hommel  0.05        0.054 0.01826667
4      AR1 5  3  0.1    LSU  0.05        0.090 0.03043333
5      AR1 5  5  0.1 Hommel  0.05        0.054 0.05400000
6      AR1 5  5  0.1    LSU  0.05        0.054 0.05400000
7      AR1 5  1 0.25 Hommel  0.05        0.054 0.01080000
8      AR1 5  1 0.25    LSU  0.05        0.054 0.01080000
9      AR1 5  3 0.25 Hommel  0.05        0.053 0.01816667
10     AR1 5  3 0.25    LSU  0.05        0.091 0.03166667
11     AR1 5  5 0.25 Hommel  0.05        0.052 0.05200000
12     AR1 5  5 0.25    LSU  0.05        0.052 0.05200000
```

This list can then be written into a file or included into a manuscript, etc.

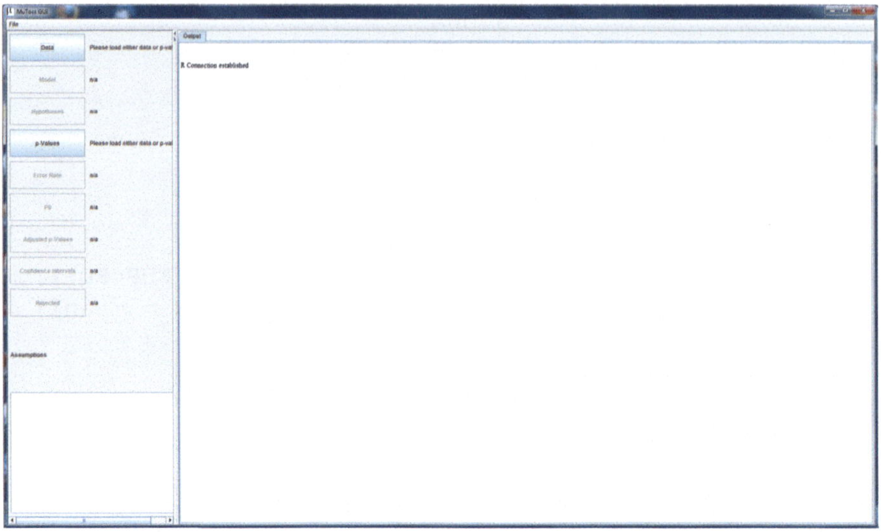

Fig. 8.1 Screenshot of the μTOSS graphical user interface when initially invoked

8.3.2 The μTOSS Graphical User Interface

The μTOSS graphical user interface is designed on the one hand to facilitate the usage of `mutoss` functions without having to program on the `R` console and on the other hand as some kind of online user's guide for finding appropriate methods for a given specification of a multiple testing problem.

To explain the latter approach, Fig. 8.1 displays a screenshot of the μTOSS graphical user interface when started by executing the `mutossGUI()` command on the `R` console. The panel of buttons on the left-hand side is meant to be traversed from top to bottom, leading the user through the entire workflow of reading in the data, defining the statistical model and the multiple test problem at hand, choosing an appropriate multiple test for the defined multiple test problem and displaying and saving the test results. Buttons subsequently only become clickable once the respectively required information has been provided. Furthermore, once the multiple test problem has been defined, all `mutoss` functions which are unsuitable for this problem are hidden from the user and can thus not be invoked by choosing a method from the displayed menu list, see Fig. 8.2.

In the example displayed in Fig. 8.2, the user loaded a family of marginal p-values into the `mutoss` system and specified that FDR control at level 0.05 is targeted. Hence, only FDR-controlling multiple tests operating on marginal p-values are listed when clicking the button labeled "Rejected".

Finally, before the chosen multiple test procedure is actually executed, a pop-up window informs the user about the procedure and its underlying assumptions and provides references to underlying publications, see Fig. 8.3. Thus, the user can check

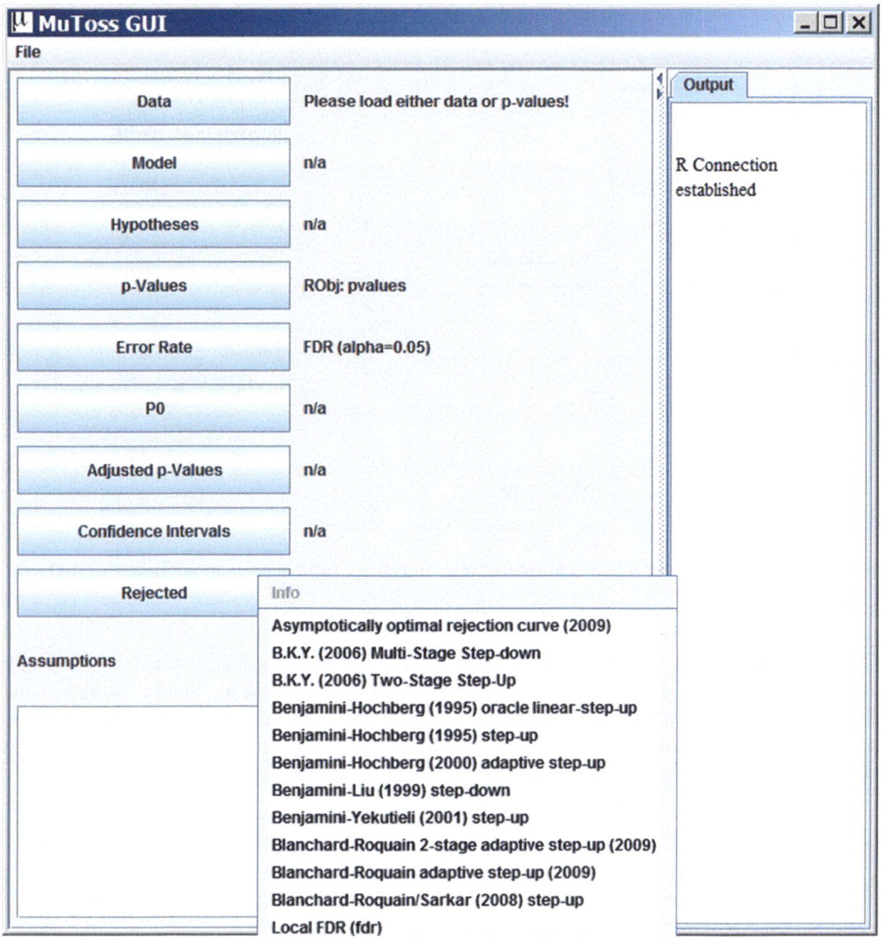

Fig. 8.2 Screenshot of the μTOSS graphical user interface: Selection of appropriate multiple test procedures

if the chosen multiple test is really appropriate and potentially choose another one if this is not the case.

The right subwindow of the μTOSS graphical user interface displays the results of the application of the chosen multiple test procedure to the user's dataset and further diagnostic plots and model parameters. We will provide examples of this in later chapters based on concrete datasets.

The implementation of `mutossGUI` is based on the `rJava` package, guaranteeing a high degree of platform independence, meaning that whenever a Java installation is present on the user's operating system, `mutossGUI` should work with a more or less unified look and feel.

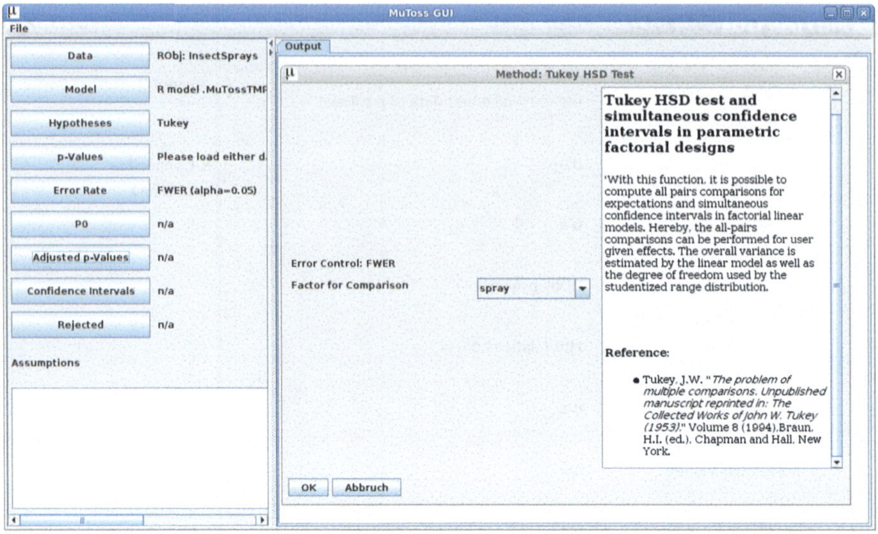

Fig. 8.3 Screenshot of the μTOSS graphical user interface: Information about Tukey's HSD test

Acknowledgments The μTOSS coding team consists of Gilles Blanchard, Niklas Hack, Frank Konietschke, Kornelius Rohmeyer, Jonathan Rosenblatt, Marsel Scheer, Wiebke Werft and the author. We gratefully acknowledge financial support by PASCAL2 and thank the machine learning / intelligent data analysis group at Berlin Institute of Technology for hosting the first major μTOSS coding phase.

References

Blanchard G, Dickhaus T, Hack N, Konietschke F, Rohmeyer K, Rosenblatt J, Scheer M, Werft W (2010) μTOSS - Multiple hypothesis testing in an open software system. J Mach Learn Res: Workshop and Conf Proc 11:12–19

Bretz F, Hothorn T, Westfall P (2010) Multiple comparisons using R.Chapman and Hall/CRC. CRC Press, Boca Raton

Dudoit S, van der Laan MJ (2008) Multiple testing procedures with applications to genomics. Springer Series in Statistics. Springer, New York

Genz A, Bretz F (2009) Computation of multivariate normal and t probabilities. Lecture notes in statistics 195. Springer, Berlin. doi:10.1007/978-3-642-01689-9

Hollander M, Wolfe DA (1973) Nonparametric statistical methods. Wiley series in probability and mathematical statistics. Wiley, New York

Hothorn T, Bretz F, Westfall P (2008) Simultaneous inference in general parametric models. Biom J 50(3):346–363

Mann HB, Whitney DR (1947) On a test of whether one of two random variables is stochastically larger than the other. Ann Math Stat 18:50–60. doi:10.1214/aoms/1177730491

Piepho HP (2004) An algorithm for a letter-based representation of all-pairwise comparisons. J Comput Graph stat 13(2):456–466

Welch B (1938) The significance of the difference between two means when the population variances are unequal. Biometrika 29:350–362

Westfall PH, Young SS (1993) Resampling-based multiple testing: examples and methods for p-value adjustment. Wiley series in probability and mathematical statistics. Applied probability and statistics. Wiley, New York

Wilcoxon F (1945) Individual comparisons by ranking methods. Biometrics Bull 1(6):80–83. doi:10.1214/aoms/1177730491

Part II
From Genotype to Phenotype

Chapter 9
Genetic Association Studies

Abstract In genetic association studies, one analyzes associations between a (potentially very large) set of genetic markers and a phenotype of interest. This is a particular multiple test problem which has several challenging aspects, for instance the high dimensionality of the statistical parameter and the discreteness of the statistical model. In this chapter, we discuss how to fine-tune multiple tests that we have described theoretically in Part I in order to address these challenges. In particular, we propose the usage of realized randomized p-values in data-adaptive multiple tests and show how linkage disequilibrium among genetic markers can be employed to construct simultaneous test procedures and to establish probability bounds which lead to effective numbers of tests. Finally, we analyze (positive) dependency properties among test statistics and the applicability of standard margin-based multiple tests. The methods are applied to two real-life datasets.

From the statistical point of view, genetic association studies lead to the problem of simultaneous categorical data analysis. Here, we focus on specific study designs in which associations between a set of bi-allelic genetic markers (typically single nucleotide polymorphisms, SNPs for short) and a dichotomous phenotype (also referred to as endpoint, typically a disease indicator) are to be analyzed in a case-control setup based on a sample of unrelated individuals. Furthermore, we assume that data have already been pre-processed, including quality control steps like for instance tests for Hardy-Weinberg equilibrium in controls, cf., among others, Finner et al. (2010).

The association analysis will be formalized mathematically by a family of tests for association in (2×2) or (2×3) contingency tables. For a detailed discussion of the appropriate choice of table layout according to genetic modeling, see, for instance, Chap. 10 in the textbook by Ziegler and König (2006). Our proposed methodology takes into account the discreteness of the test problem, the dependency structure among genetic markers which can technically be described by linkage disequilibrium

T. Dickhaus, *Simultaneous Statistical Inference*,
DOI: 10.1007/978-3-642-45182-9_9,

(LD) matrices, and the fact that (particularly in replication studies) the existence of a non-negligible proportion of false null hypotheses can be assumed.

9.1 Statistical Modeling and Test Statistics

In what follows, m denotes the number of considered markers. In this, markers can be both genotyped (observed) or imputed (i.e., estimated using population genetics techniques and a priori information from a reference population, see Marchini et al. (2007), Willer et al. (2008)). Imputed marker genotypes usually have a very high degree of certainty, so that they are widely considered as regular observed genotypes (cf. Howie et al. (2009), Li et al. (2012), The 1000 Genomes Consortium (2010)). Hence, we will not distinguish between actually observed and imputed genotypes.

The data for one specific genetic locus can be organized in a contingency table. Let us assume that the two rows of this table correspond to the phenotype and its (two or three) columns contain the marker counts. Since we want to treat the cases of (2×2) and (2×3) tables simultaneously all along the way, we will denote by $\mathbf{n}$ the vector containing all the marginals of the table. Therefore, $\mathbf{n}$ can have different dimensionality depending on the context. In the (2×2) table case, we have $\mathbf{n} = (n_{1.}, n_{2.}, n_{.1}, n_{.2}) \in \mathbb{N}^4$ while we have $\mathbf{n} = (n_{1.}, n_{2.}, n_{.1}, n_{.2}, n_{.3}) \in \mathbb{N}^5$ in the (2×3) table case. In both cases, we define the number of observational units by $N = n_{1.} + n_{2.}$. In the case of a (2×3) table, N is therefore equal to the number of individuals in the study, while it equals the number of alleles (twice the number of study participants) in the case of a (2×2) table. Accordingly, an observed table $\mathbf{x}$ (say) takes the form $\mathbf{x} = \begin{pmatrix} x_{11} & x_{12} \\ x_{21} & x_{22} \end{pmatrix} \in \mathbb{N}^{2\times 2}$ in case of a (2×2) table and $\mathbf{x} = \begin{pmatrix} x_{11} & x_{12} & x_{13} \\ x_{21} & x_{22} & x_{23} \end{pmatrix} \in \mathbb{N}^{2\times 3}$ in the (2×3) case, see Table 9.1 for a schematic representation.

Aim of the analysis is to test the null hypothesis H of no association of phenotype and the genetic marker corresponding to $\mathbf{x}$ against its (two-sided) alternative hypothesis K that phenotype and marker are associated.

The conditional probability of observing $\mathbf{x}$ given $\mathbf{n}$ under H will be denoted by $f(\mathbf{x}|\mathbf{n})$ and is (in a compact, self-explaining notation) given by

Table 9.1 Schematic representation of data for an association test problem at one genetic locus, where the two possible alleles are denoted by A_1 and A_2

Genotype/Allele	A_1A_1	A_1A_2	A_2A_2	Σ	A_1	A_2	Σ
Phenotype 1	x_{11}	x_{12}	x_{13}	$n_{1.}$	x_{11}	x_{12}	$n_{1.}$
Phenotype 0	x_{21}	x_{22}	x_{23}	$n_{2.}$	x_{21}	x_{22}	$n_{2.}$
Absolute count	$n_{.1}$	$n_{.2}$	$n_{.3}$	N	$n_{.1}$	$n_{.2}$	N

The left part of the table corresponds to a genotypic association test problem and the right part to an allelic association test problem

$$f(\mathbf{x}|\mathbf{n}) = \frac{\prod_{n \in \mathbf{n}} n!}{N! \prod_{x \in \mathbf{x}} x!}. \tag{9.1}$$

One classical chi-square statistic for assessing association of the phenotype and the genetic marker from the observed data $\mathbf{x}$ is given by

$$Q_{\text{assoc.}}(\mathbf{x}) = \sum_r \sum_c \frac{(x_{rc} - e_{rc})^2}{e_{rc}}, \tag{9.2}$$

where r runs over the rows and c over the columns of $\mathbf{x}$. In (9.2), the numbers $e_{rc} = n_{r.}n_{.c}/N$ denote the (under H) expected cell counts given N and the marginal counts contained in $\mathbf{n}$. The statistic $Q_{\text{assoc.}}$ is commonly referred to as Pearson's chi-squared statistic for association.

Notice that the exact (non-asymptotic) distribution of $Q_{\text{assoc.}}(\mathbf{X})$, conditional to $\mathbf{n}$, is implied by (9.1). Among others, Weir (1996) and Wigginton et al. (2005) have proposed to utilize this conditional distribution for the calculation of an "exact" p-value which is given by

$$p_{\text{assoc.}}(\mathbf{x}) = \sum_{\tilde{\mathbf{x}}} f(\tilde{\mathbf{x}}|\mathbf{n}), \tag{9.3}$$

where the summation is carried out over all tables $\tilde{\mathbf{x}}$ with marginals $\mathbf{n}$ for which $Q_{\text{assoc.}}(\tilde{\mathbf{x}}) \geq Q_{\text{assoc.}}(\mathbf{x})$.

However, for the calculation of effective numbers of tests based on the LD structure in the target population (see Sect. 9.3 below), unconditional asymptotic distributions are more tractable. Under the null hypothesis H, the asymptotic ($N \to \infty$) distribution of $Q_{\text{assoc.}}(\mathbf{X})$ is chi-squared with ν degrees of freedom, where $\nu = 1$ for the (2×2) table case and $\nu = 2$ for the (2×3) table case. Hence, an asymptotic p-value can be calculated as $\tilde{p}_{\text{assoc.}}(\mathbf{x}) = 1 - F_{\chi^2_\nu}(Q_{\text{assoc.}}(\mathbf{x}))$.

In the case of a genotypic association test problem (where the three possible allele pairs are of interest), often trend tests are considered instead of mere association tests. The corresponding (Cochran-Armitage) trend test statistic Q_{trend} is given by

$$Q_{\text{trend}}(\mathbf{x}) = \frac{\left[\sum_{c=1}^{3} x_{1c}(w_c - \bar{w})\right]^2}{p(1-p)\sum_{c=1}^{3} n_{.c}(w_c - \bar{w})^2}. \tag{9.4}$$

In (9.4), $p = n_{1.}/N$ denotes the relative frequency of cases, the weights $(w_c : 1 \leq c \leq 3)$ are used to quantitatively express the influence of the occurrence of the risk allele on the disease risk and $\bar{w} = \sum_{c=1}^{3} n_{.c}w_c/N$. It is worth noticing that $Q_{\text{trend}}(\mathbf{x}) = Nr^2_{wy}$, where r_{wy} is Pearson's correlation coefficient of the two vectors w and y, each of length N. Each element in w and y, respectively, corresponds to one observational unit. The vector $w \in \mathbb{R}^N$ contains the weights associated with the cells (columns) to which the observational units belong and the elements in $y \in \{0, 1\}^N$ are disease indicators, so that $y_i = 1$ if and only if observational unit i corresponds to a case. Since, by definition of $\bar{w}$, $\sum_{c=1}^{3} x_{1c}(w_c - \bar{w}) = -\sum_{c=1}^{3} x_{2c}(w_c - \bar{w})$, the value $Q_{\text{trend}}(\mathbf{x})$ is invariant to the coding with respect to the disease status. Furthermore, Q_{trend} is the score statistic for testing the hypothesis $\{\beta = 0\}$ in the model $\pi_c = \alpha + \beta w_c$, $1 \leq c \leq 3$, as noted by Agresti (2002), p. 182. In this, π_c denotes the probability that a randomly chosen individual from the target population is positive with respect to the target event (i.e., is a case), given that his/her genotype corresponds to column c. Popular choices for the weights are $w = (w_1, w_2, w_3)^\top = (0, 1, 2)^\top$ (coding an additive risk allele contribution) or $w_1 = \sum_{k=1}^{n_{.1}} k/n_{.1}$, $w_2 = \sum_{k=n_{.1}+1}^{n_{.1}+n_{.2}} k/n_{.2}$, $w_3 = \sum_{k=n_{.1}+n_{.2}+1}^{N} k/n_{.3}$ (average rank scores).

In total analogy to (9.3), an exact p-value corresponding to Q_{trend} can be computed. Moreover, the asymptotic distribution of $Q_{\text{trend}}(\mathbf{X})$ is chi-squared with one degree of freedom under H.

The two statistics $Q_{\text{assoc.}}$ and Q_{trend} are very widely considered in practice, but they are by far not the only possible choices. Several other test statistics and p-values for the association test problem are discussed in Chap. 3 of Zheng et al. (2012), by Langaas and Bakke (2013) and by Dickhaus et al. (2012); see also the references therein.

To keep notation feasible, we restricted our attention up to now to one specific genetic locus. However, for the subsequent discussion it is necessary to introduce notation for the practically relevant case that $m > 1$ markers are simultaneously under investigation. To this end, Definition 9.1 extends the sampling model to this case.

Definition 9.1. Consider a statistical model $(\mathscr{X}, \mathscr{F}, (\mathbb{P}_\vartheta)_{\vartheta \in \Theta})$ which can be decomposed into local statistical models $(\mathscr{X}_j, \mathscr{F}_j, (\mathbb{P}_{\vartheta_j})_{\vartheta_j \in \Theta_j})_{1 \leq j \leq m}$, such that

$$\mathscr{X} = \bigtimes_{j=1}^{m} \mathscr{X}_j,\ \mathscr{F} = \bigotimes_{j=1}^{m} \mathscr{F}_j,\ \Theta = \bigtimes_{j=1}^{m} \Theta_j,\ \mathbb{P}_{\vartheta_j}(A_j) = \mathbb{P}_\vartheta(\pi_j^{-1}(A_j)) \text{ for } A_j \in \mathscr{F}_j,$$

where $\pi_j : (\mathscr{X}, \mathscr{F}) \to (\mathscr{X}_j, \mathscr{F}_j)$ denotes the projection on the j-th coordinate.

(a) Assume that for all $1 \leq j \leq m$, $\mathscr{X}_j = \mathbb{N}^{2\times 2}$ and $\mathscr{F}_j = 2^{\mathscr{X}_j}$. An observation $\mathbf{x}_j = \begin{pmatrix} x_{11}^{(j)} & x_{12}^{(j)} \\ x_{21}^{(j)} & x_{22}^{(j)} \end{pmatrix} \in \mathscr{X}_j$ necessarily fulfills $x_{11}^{(j)} + x_{12}^{(j)} = n_{1.}$ and $x_{21}^{(j)} + x_{22}^{(j)} = n_{2.}$ by experimental design. Denoting the multinomial distribution with c categories, sample size n and vector of probabilities p by $\mathscr{M}_c(n, p)$, we have that, for every j, the pair of random variables $(X_{11}^{(j)}, X_{12}^{(j)})$ is distributed as $\mathscr{M}_2(n_{1.}, p_j)$, with $p_j = (p_{1j}, p_{2j})^\top$ taking the role of ϑ_j in our general setup. We are considered with the point null hypothesis $H_j : p_j = P_j$, where $P_j = (P_{1j}, P_{2j})^\top$ denotes the vector of (expected) allele frequencies at position j in the entire target population (which is unknown in practice). Canonical estimators for P_{1j} and P_{2j} are given by $\hat{p}_{1j} = n_{.1}^{(j)}/N$, $\hat{p}_{2j} = n_{.2}^{(j)}/N$, where $n_{.1}^{(j)} = x_{11}^{(j)} + x_{21}^{(j)}$ and $n_{.2}^{(j)} = x_{12}^{(j)} + x_{22}^{(j)}$ denote the column counts in the j-th contingency table. We refer to the resulting multiple test problem as a multiple allelic association test problem.

(b) Assume that for all $1 \leq j \leq m$, $\mathscr{X}_j = \mathbb{N}^{2\times 3}$ and $\mathscr{F}_j = 2^{\mathscr{X}_j}$. An observation $\mathbf{x}_j = \begin{pmatrix} x_{11}^{(j)} & x_{12}^{(j)} & x_{13}^{(j)} \\ x_{21}^{(j)} & x_{22}^{(j)} & x_{23}^{(j)} \end{pmatrix} \in \mathscr{X}_j$ again fulfills $x_{11}^{(j)} + x_{12}^{(j)} + x_{13}^{(j)} = n_{1.}$ and $x_{21}^{(j)} + x_{22}^{(j)} + x_{23}^{(j)} = n_{2.}$ by experimental design. For every j, the triple of random variables $(X_{11}^{(j)}, X_{12}^{(j)}, X_{13}^{(j)})$ is distributed as $\mathscr{M}_3(n_{1.}, p_j)$, with unknown parameter vector $p_j = (p_{1j}, p_{2j}, p_{3j})^\top$. The point hypothesis H_j that we are concerned with is then given by $H_j : p_j = P_j = (P_{1j}, P_{2j}, P_{3j})^\top$, where P_j denotes the vector of expected genotype frequencies at position j in the entire target population, in analogy to part (a). We let $n_{.1}^{(j)} = x_{11}^{(j)} + x_{21}^{(j)}$, $n_{.2}^{(j)} = x_{12}^{(j)} + x_{22}^{(j)}$, $n_{.3}^{(j)} = x_{13}^{(j)} + x_{23}^{(j)}$, and $\hat{p}_{ij} = n_{.i}^{(j)}/N$, $i = 1, 2, 3$, and refer to the resulting multiple test problem as a multiple genotypic association test problem.

Notice that Definition 9.1 suggests to re-formulate the association test problem at locus j as a goodness-of-fit test problem, where the empirical distribution of the observed genotype frequencies in cases is compared with the vector $(\hat{p}_{ij} : i \geq 1)$ of relative column counts. In practice, in particular for large sample sizes N and $n_{2.}/N$ not too small, this re-formulation will typically not make a decisive difference, but it facilitates the mathematical treatment of the multiple association test problems, as remarked by Moskvina and Schmidt (2008), cf. Appendix A in their paper. To illustrate this, consider the (2×2) table case. Let $\pi = (\pi_1, \pi_2)^\top$ with $\pi_2 = 1 - \pi_1$ denote the vector containing the (true unknown) probabilities for cases to exhibit the alleles corresponding to columns 1 or 2, respectively. Then, the statistic

$$Q_{\text{Pearson}}(\mathbf{X}^{(j)}) = \frac{(X_{11}^{(j)} - n_{1.}\hat{p}_{1j})^2}{n_{1.}\hat{p}_{1j}} + \frac{(X_{12}^{(j)} - n_{1.}\hat{p}_{2j})^2}{n_{1.}\hat{p}_{2j}} = \frac{(X_{11}^{(j)} - n_{1.}\hat{p}_{1j})^2}{n_{1.}\hat{p}_{1j}\hat{p}_{2j}} \tag{9.5}$$

is formally identical to Pearson's chi-squared statistic for goodness-of-fit. It is asymptotically equivalent to the likelihood ratio statistic for the point hypothesis $\pi = (\hat{p}_{1j}, \hat{p}_{2j})^\top$ (see, e.g., Spokoiny and Dickhaus (2014), Sect. 6.3). While $Q_{\text{assoc.}}$ involves four summands, Q_{Pearson} only involves two, making the mathematical analysis much simpler, especially with respect to the (asymptotic) correlation structure among different chi-square statistics corresponding to different genomic positions. Notice that $Q_{\text{assoc.}}(\mathbf{x}^{(j)}) = NQ_{\text{Pearson}}(\mathbf{x}^{(j)})/n_{2.}$, both in the (2×2) table case and in the (2×3) table case.

9.2 Estimation of the Proportion of Informative Loci

Often, genetic association studies are planned to consist of two stages: A screening stage and a validation stage, with independent data. From the statistical perspective, this two-stage approach has already been described in detail by Wasserman and Roeder (2009) and Meinshausen et al. (2009), cf. Sect. 7.3.2. For instance, one may apply a screening criterion (for example, FDR control at level $1/2$ as applied by Dickhaus et al. (2012)) to the data from the first stage. Only markers that are selected by a multiple test employing this screening criterion are considered in the second (validation) stage. This may be considered as some kind of model selection. Then, in the validation stage, the total number m of (remaining) hypotheses will typically be much smaller than the number of markers that have initially been considered. Furthermore, the proportion π_0 of true hypotheses among the m (remaining) ones will also be much smaller than 1, making the usage of data-adaptive multiple tests relying on a pre-estimation of π_0 a natural choice; cf. Sect. 3.1.3.

However, if this strategy is pursued in connection with exact (conditional) p-values as defined in (9.3), it is recommendable to transform these p-values into realized randomized p-values according to Theorem 2.3. Let us illustrate the advantage of realized randomized p-values in this context by a real-data example. The underlying dataset has been generated by the study reported by The Wellcome Trust Case Control Consortium (2007). Here, we restrict attention to the substudy for Crohn's disease and mimic the situation of a replication study. To this end, we first split the full dataset comprising $N = 4{,}688$ individuals into two halves of size $N/2$ each. To the first split, we apply a screening for candidate loci by applying the linear step-up test from Definition 5.6 at FDR level $1/2$. This leads to 1,778 rejections (candidates). Then, in the second split, we calculate p-values only for these pre-screened positions. Fig. 9.1 displays ecdfs of realized randomized and nonrandomized p-values corresponding to $Q_{\text{assoc.}}$ for the $m = 175$ pre-screened loci which are on chromosome 1.

The advantage of working with realized randomized p-values (solid curve in Fig. 9.1) can clearly be observed. For the standard choice $\lambda = 1/2$ for the Schweder-Spjøtvoll estimator cf. (3.2), we obtain $\hat{\pi}_0^{\text{rand.}}(1/2) = 0.777$ in the case that we utilize realized randomized p-values in this estimation procedure, while utilization of the non-randomized counterparts results in $\hat{\pi}_0^{\text{non-rand.}}(1/2) = 0.937$.

Qualitatively, the same behavior of the ecdfs of randomized and non-randomized p-values can be observed if all 1,778 pre-screened loci on autosomes are analyzed together, see Fig. 9.1 of Dickhaus et al. (2012). A counter-argument against the strategy of estimating π_0 by utilizing realized randomized p-values may be that results depend on pseudo random numbers and are therefore not entirely reproducible. If fully reproducible results are wanted, it is possible to replace $\hat{\pi}_0^{\text{rand.}}$ by its conditional expectation with respect to randomization, as described in Appendix III of Dickhaus et al. (2012).

9.3 Effective Numbers of Tests via Linkage Disequilibrium

The underlying biological (inheritance) mechanism entails strong (positive) correlations between chi-square statistics corresponding to loci which are in linkage disequilibrium with each other. Linkage disequilibrium is the technical way to refer to correlations between the allelic states of different genetic markers in the same

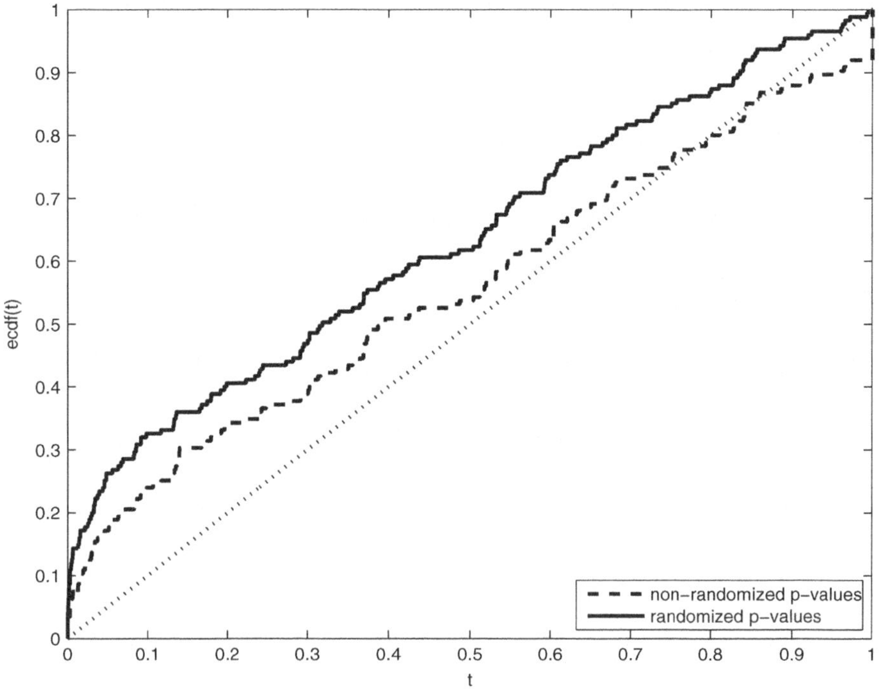

Fig. 9.1 Empirical cumulative distribution functions of realized randomized and nonrandomized p-values corresponding to $Q_{\mathrm{assoc.}}$. Data are taken from the Crohn's disease substudy of The Wellcome Trust Case Control Consortium (2007), Chromosome 1

chromosome, see Lewontin and Kojima (1960). In human populations some combinations of alleles along the same chromosome (haplotypes) occur at frequencies that are different from what would be expected out of random combinations of the markers' allelic frequencies. As discussed in Chap. 4, strong positive correlations among test statistics help to reduce the "effective multiplicity" of the multiple test problem.

The following results are taken from Dickhaus and Stange (2013). They refer to the notation developed in Definition 9.1 and provide a stochastic representation for the asymptotic distribution of the vector of m chi-square statistics for association at m genomic positions under the global hypothesis H_0, both in the multiple allelic association test model (Lemma 9.1) and in the multiple genotypic association test model (Lemma 9.2). With these asymptotic distributions at hand, a simultaneous test procedure for FWER control can precisely be calibrated, at least for large sample sizes N. We may remark here that Lemma 4.1 applies, entailing that the resulting simultaneous test procedure controls the FWER strongly (at least asymptotically).

Lemma 9.1. *Consider the multiple allelic association test model with* m (2×2)-*contingency tables. Let* $Z = (Z_1, \ldots, Z_m)^\top$, *where, for all* $1 \leq j \leq m$,

$$Z_j = \sqrt{\frac{N}{n_{2.}}} \frac{X_{11}^{(j)} - n_{1.}\hat{p}_{1j}}{\sqrt{n_{1.}\hat{p}_{1j}\hat{p}_{2j}}}.$$

Then, the distribution of Z converges weakly to $\mathcal{N}_m(0, \Sigma)$ under the global hypothesis H_0 as $N \to \infty$. In this, $\Sigma_{ij} = \rho_{ij}$ (say), and ρ_{ij} is equal to Pearson's haplotypic correlation coefficient of markers i and j (which is referred to as linkage disequilibrium coefficient in the genetics literature and is tabulated for several target populations). Notice that $Q_{assoc.}(\mathbf{X}^{(j)}) = Z_j^2$, $1 \leq j \leq m$.

Lemma 9.2. *Consider the multiple genotypic association test model with m (2×3)-contingency tables. Let, for $1 \leq j \leq m$,*

$$Z_{1,j} = \frac{X_{11}^{(j)} - n_{1.}P_{1j}}{\sqrt{n_{1.}P_{1j}(1 - P_{1j})}}, \tag{9.6}$$

$$Z_{2,j} = \frac{P_{2j}(X_{11}^{(j)} - n_{1.}P_{1j}) + (1 - P_{1j})(X_{12}^{(j)} - n_{1.}P_{2j})}{\sqrt{n_{1.}P_{2j}(1 - P_{1j})(1 - P_{1j} - P_{2j})}}. \tag{9.7}$$

Then, for $N \to \infty$, $(Z_{1,j}, Z_{2,j})^\top$ converges under H_j in distribution to $(Z_1, Z_2)^\top$ with $(Z_1, Z_2)^\top \sim \mathcal{N}_2(0, E_2)$, the standard normal distribution on $\mathbb{R}^2$. Furthermore, $Q_{assoc.}(\mathbf{X}^{(j)})$ converges in distribution to $Z_1^2 + Z_2^2$. Finally, under the global hypothesis H_0, it holds for all $1 \leq j, k \leq m$: For any tuple $(\ell_1, \ell_2) \in \{1, 2\}^2$, the joint distribution of $(Z_{\ell_1,j}, Z_{\ell_2,k})^\top$ converges weakly to a bivariate normal distribution with correlation coefficient given by

$$\lim_{N \to \infty} Cov(Z_{\ell_1,j}, Z_{\ell_2,k}) = r_{j,k}(\ell_1, \ell_2) \text{ (say)}. \tag{9.8}$$

Consequently, the vector $\mathbf{Q}_{assoc.}(\mathbf{X}) = (Q_{assoc.}(\mathbf{X}^{(1)}), \ldots, Q_{assoc.}(\mathbf{X}^{(m)}))^\top$ asymptotically follows a (generalized) multivariate central chi-squared distribution under H_0, with correlation structure given by

$$\lim_{N \to \infty} Cov(Q_{assoc.}(\mathbf{X}^{(j)}), Q_{assoc.}(\mathbf{X}^{(k)})) = 2 \sum_{\ell_1=1}^{2} \sum_{\ell_2=1}^{2} r_{j,k}^2(\ell_1, \ell_2). \tag{9.9}$$

The correlations $r_{j,k}(\ell_1, \ell_2)$ in (9.8) only depend on the expected genotype frequencies P_{ij}, P_{ik}, $i = 1, 2, 3$ and on the second-order joint probabilities of genotype pairs.

The results of Lemmas 9.1 and 9.2 can straightforwardly be used to calculate effective numbers of tests based on probability bounds as described in Sect. 4.3. To give numerical examples, Dickhaus et al. (2012) applied the product-type probability bound β_2 from (4.14) to the Crohn's disease sub-dataset from The Wellcome Trust Case Control Consortium (2007) and obtained based on Lemma 9.1 an effective number of 329,079.66 association tests, whereas the total number of markers under consideration in this sub-study is (after quality control) equal to $m = 455{,}086$. Application of the same bound to the small-scale replication study reported by Herder et al. (2008), comprising $m = 44$ SNPs on ten different genes, led to an effective number of 16.73 tests. The explicit numerical formula for the effective number of tests based on Lemma 9.1 in connection with β_2 from (4.14) has originally been derived by Moskvina and Schmidt (2008), see part (ii) of Example 4.1.

9.4 Combining Effective Numbers of Tests and Pre-estimation of π_0

Based on the considerations in Sects. 9.2 and 9.3, it is near at hand to construct a simultaneous test procedure that takes into account both the effective number of tests and the estimated proportion of informative loci. Along these lines, Dickhaus et al. (2012) proposed the following algorithm.

Algorithm 9.1

1. *For $j = 1, \ldots, m$, build the contingency table $\mathbf{x}_j$ carrying the information gathered for association of marker j and the phenotype under investigation.*
2. *For $j = 1, \ldots, m$, compute the realized randomized p-value $p^{rand.}(\mathbf{x}_j, u_j)$ and the non-randomized version $p(\mathbf{x}_j)$ (say) by making use of one of the testing strategies described in Sect. 9.1 and the realization u_j of an UNI$[0,1]$-distributed random variable which is stochastically independent of $\mathbf{X}_j$.*
3. *Compute $\hat{\pi}_0(\lambda)$ by calculating the ecdf. of $(p^{rand.}(\mathbf{x}_j, u_j), j = 1, \ldots, m)$. In practice, it is convenient to use the value $\lambda = 0.5$ for the tuning parameter.*
4. *Determine the effective number of tests by utilizing correlation values obtained from an appropriate LD matrix of the m markers. Denote the resulting (estimated) effective number of tests by Eff.*
5. *For a pre-defined FWER level α, determine the list of associated markers by performing the multiple test $\varphi = (\varphi_j, j = 1, \ldots, m)$, where $\varphi_j(\mathbf{x}_j) = \mathbf{1}_{p(\mathbf{x}_j) \leq t^*}$ with $t^* = \alpha / (Eff \cdot \hat{\pi}_0(\lambda))$.*

Remark 9.1.

(a) Notice that the realized randomized p-values are only used in the third step of Algorithm 9.1 for estimation of π_0, while for final decision making in the fifth step the non-randomized p-values are used. This ensures accurate estimation of π_0 on the one hand and reproducibility of the test result on the other hand.
(b) The underlying assumption of Algorithm 9.1 is that the pairwise marker correlations are on average of not smaller magnitude in the group of markers which are not associated with the phenotype under investigation than in the group of informative markers. This assumption can be formalized as the relationship

$$\pi_0 = \frac{M_0}{M} \geq \frac{\mathrm{Eff}(I_0)}{\mathrm{Eff}} \quad \text{or, equivalently,} \quad \pi_0 \mathrm{Eff} \geq \mathrm{Eff}(I_0), \tag{9.10}$$

where $\text{Eff}(I_0)$ denotes the effective number of tests within the subset of markers for which the null hypothesis of no association with the phenotype holds. Of course, assumption (9.10) cannot be verified in practice, because I_0 is unobservable. However, it appears natural, because informative markers are assumed to be sparsely distributed among the genome and consequently most of their pairwise LD values should be of low magnitude. Non-associated markers (with the phenotype), however, lie dense and should have on average a higher pairwise correlation.

By means of computer simulations, Dickhaus et al. (2012) demonstrated that Algorithm 9.1 keeps the FWER accurately and outperforms previous methods, which only make use of the LD structure or the estimated proportion of true/false null hypotheses, in terms of multiple power.

9.5 Applicability of Margin-Based Methods

Although LD databases like the one by The International HapMap Consortium (2005) become more and more reliable and technical developments allow for larger and larger studies such that LD matrices may be estimated from the data in controls of the actual study at hand, one may nevertheless want to resort on margin-based multiple tests which do not utilize LD information explicitly, in particular because standard implementations make their application very easy in practice. Especially when the number m of considered markers is very large, the asymptotic p-values based on chi-square approximations are convenient to use in SUD tests, because they can be computed very fast. In view of Lemma 4.2, one resulting question of interest is if for instance the multiple test φ^{Hommel} from Definition 5.4 or the linear step-up test φ^{LSU} from Definition 5.6 control the FWER or the FDR, respectively, for such p-values. These tests have guaranteed control of the FWER or the FDR, respectively, if the p-values exhibit higher-order positive dependency properties like PRDS or MTP_2, cf. Sect. 5.1. Moreover, due to Lemma 3.1, they have higher multiple power than their counterparts for the totally generic case of arbitrary dependency structure among p-values. Unfortunately, however, pairwise positive correlations are not sufficient for proving PRDS or MTP_2, see e.g. Example 3.2. of Karlin and Rinott (1980). Hence, at least assumptions on the structure of the underlying LD matrix and maybe also on their actual entries have to be imposed for these concepts to hold true. For this reason, Dickhaus (2012) carried out an extensive simulation study of the FWER and FDR behavior, respectively, of φ^{Hommel} and φ^{LSU} for various cases with multivariate chi-square distributed test statistics under null hypotheses. It turned out that φ^{Hommel} and φ^{LSU} keep the respective type I error rate accurately, for several correlation matrices R as in Definition 4.6. These results indicate that the answer to the question of applicability of these multiple tests to asymptotic p-values originating from genetic association analyses is positive. In particular for screening purposes based on the FDR criterion, φ^{LSU} appears to be an appropriate choice.

Acknowledgments This chapter makes use of data generated by the Wellcome Trust Case Control Consortium. A full list of the investigators who contributed to the generation of the data is available from http://www.wtccc.org.uk. Funding for the Wellcome Trust Case Control Consortium project was provided by the Wellcome Trust under award 076113. Parts of this chapter originated from joint work with Klaus Straßburger, Daniel Schunk, Carlos Morcillo-Suarez, Thomas Illig, Arcadi Navarro and Jens Stange. I am grateful to Mette Langaas and Øyvind Bakke for inviting me and for their hospitality during my visit to Norwegian University of Science and Technology (NTNU), for many fruitful discussions and for some valuable references.

References

Agresti A (2002) Categorical data analysis. Wiley Series in Probability and Mathematical Statistics, 2nd edn. Wiley, Chichester

Dickhaus T (2012) Simultaneous Statistical Inference in dynamic factor models. SFB 649 Discussion Paper 2012–033, Sonderforschungsbereich 649, Humboldt Universität zu Berlin, Germany. http://sfb649.wiwi.hu-berlin.de/papers/pdf/SFB649DP2012-033.pdf

Dickhaus T, Stange J (2013) Multiple point hypothesis test problems and effective numbers of tests for control of the family-wise error rate. Calcutta Statis Assoc Bull, to appear

Dickhaus T, Strassburger K, Schunk D, Morcillo-Suarez C, Illig T, Navarro A (2012) How to analyze many contingency tables simultaneously in genetic association studies. Stat Appl Genet Mol Biol 11(4):Article 12

Finner H, Straßburger K, Heid IM, Herder C, Rathmann W, Giani G, Dickhaus T, Lichtner P, Meitinger T, Wichmann HE, Illig T, Gieger C (2010) How to link call rate and p-values for Hardy-Weinberg equilibrium as measures of genome-wide SNP data quality. Stat Med 29(22):2347–2358

Herder C, Rathmann W, Strassburger K, Finner H, Grallert H, Huth C, Meisinger C, Gieger C, Martin S, Giani G, Scherbaum WA, Wichmann HE, Illig T (2008) Variants of the PPARG, IGF2BP2, CDKAL1, HHEX, and TCF7L2 genes confer risk of type 2 diabetes independently of BMI in the German KORA studies. Horm Metab Res 40:722–726

Howie BN, Donnelly P, Marchini J (2009) A flexible and accurate genotype imputation method for the next generation of genome-wide association studies. PLoS Genet 5:e1000,529

Karlin S, Rinott Y (1980) Classes of orderings of measures and related correlation inequalities I. Multivariate totally positive distributions. J Multivariate Anal 10:467–498

Langaas M, Bakke Ø (2013) Robust Methods for Disease-Genotype Association in Genetic Association Studies: Calculate p-values using exact conditional enumeration instead of asymptotic approximations. arXiv:1307.7536v1

Lewontin RC, Kojima KI (1960) The evolutionary dynamics of complex polymorphisms. Evolution 14:458–472

Li Y, Willer CJ, Ding J, Scheet P, Abecasis GR (2010) MaCH: using sequence and genotype data to estimate haplotypes and unobserved genotypes. Genet Epidemiol 34:816–834

Marchini J, Howie B, Myers S, McVean G, Donnelly P (2007) A new multipoint method for genome-wide association studies by imputation of genotypes. Nat Genet 39:906–913

Meinshausen N, Meier L, Bühlmann P (2009) p-Values for high-dimensional regression. J Am Stat Assoc 104(488):1671–1681. doi:10.1198/jasa.2009.tm08647

Moskvina V, Schmidt KM (2008) On multiple-testing correction in genome-wide association studies. Genet Epidemiol 32:567–573

Spokoiny V, Dickhaus T (2014) Basics of modern parametric statistics. Springer, Heidelberg, forthcoming

The 1000 Genomes Consortium (2010) A map of human genome variation from population-scale sequencing. Nature 467:1061–1073

The International HapMap Consortium (2005) A haplotype map of the human genome. Nature 437(7063):1299–1320

The Wellcome Trust Case Control Consortium (2007) Genome-wide association study of 14,000 cases of seven common diseases and 3,000 hared controls. Nature 447(7):661–678

Wasserman L, Roeder K (2009) High-dimensional variable selection. Ann Stat 37(5A):2178–2201

Weir BS (1996) Genetic data analysis II. Sinauer Associates, Sunderland, MA

Wigginton JE, Cutler DJ, Abecasis GR (2005) A Note on Exact Tests of Hardy-Weinberg Equilibrium. Am J Hum Genet 76:887–893

Willer CJ, Sanna S, Jackson AU, Scuteri A, Bonnycastle LL, Clarke R, Heath SC, Timpson NJ, Najjar SS, Stringham HM, Strait J, Duren WL, Maschio A, Busonero F, Mulas A, Albai G, Swift AJ, Morken MA, Narisu N, Bennett D, Parish S, Shen H, Galan P, Meneton P, Hercberg S, Zelenika D, Chen WM, Li Y, Scott LJ, Scheet PA, Sundvall J, Watanabe RM, Nagaraja R, Ebrahim S, Lawlor DA, Ben-Shlomo Y, Davey-Smith G, Shuldiner AR, Collins R, Bergman RN, Uda M, Tuomilehto J, Cao A, Collins FS, Lakatta E, Lathrop GM, Boehnke M, Schlessinger D, Mohlke KL, Abecasis GR (2008) Newly identified loci that influence lipid concentrations and risk of coronary artery disease. Nat Genet 40:161–169

Zheng G, Yang Y, Zhu X, Elston RC (2012) Analysis of genetic association studies. Statistics for biology and health. Springer, New York. doi:10.1007/978-1-4614-2245-7

Ziegler A, König IR (2006) A statistical approach to genetic epidemiology. Wiley, Weinheim

Chapter 10
Gene Expression Analyses

Abstract This chapter is considered with multiple tests for differential gene expression. We first formalize this multiple test problem for a variety of different study designs, including two-group models with unrelated individuals and paired groups models. Then, we analyze positive dependency properties among test statistics. The proposed multiple tests are applied to two real-life datasets from cancer research by making use of the μTOSS software. Regularized estimators for the full high-dimensional model parameter and statistical machine learning models for analyzing differential gene expression are discussed. Finally, we review some methods for incorporating functional meta information and gene set structures into the statistical methodology.

Multiple testing for differential gene expression is one of *the* prototypical examples of a two-sample problem with multiple endpoints. The aim of the statistical analysis is to find out which genes from a (typically large) set of m genes is (on average) differentially expressed between two groups (for instance, cases and controls or cancer tissue and healthy tissue from the same patients). This life science application had and continues to have an enormous influence on the development of modern multiple testing methods. It is impossible to provide a full account of all or even of all major approaches that have been pursued in the literature. In this chapter, we therefore restrict our attention mainly to easy-to-implement methods which allow for the application of inferential theory that we have developed in earlier chapters.

10.1 Marginal Models and p-values

As already mentioned in Chap. 9, an appropriate pre-processing of genetic data is an essential prerequisite for valid data analysis. However, this is not primarily in the focus of the present work. Here, we assume that data have been preprocessed and normalized before inference is initiated. In Definition 10.1 below, the observables "will usually be log-ratios for two-color data or log-intensities for single channel data,

T. Dickhaus, *Simultaneous Statistical Inference*,
DOI: 10.1007/978-3-642-45182-9_10,

although other transformations are possible" (Smyth 2004). For instance, cube-root transformations were employed by Tusher et al. (2001). Furthermore, we consider the simplest possible linear model in Definition 10.1, where just a two-group comparison per gene is the aim of the statistical analysis. More sophisticated linear models, for instance incorporating adjustments for covariates, can also be considered without changing the essential argumentation.

Definition 10.1 We consider the sample space $\mathscr{X} = \mathbb{R}^{m \times n}$, equipped with its Borel σ-field. The data-generating mechanism is mathematically represented by the random matrix $X = (X_{ij})$, where $1 \leq i \leq m$ indexes genes and $1 \leq j \leq n$ indexes the observational units (which are assumed to be the same for each gene). Assume that observational units $1, \ldots, n_1$ belong to the first group and observational units $n_1 + 1, \ldots, n$ to the second group, and let $n_2 = n - n_1$.

(a) Assume that all n observational units correspond to unrelated individuals, such that $X_1, \ldots, X_n$ can be assumed to be stochastically independent random vectors, where X_j denotes the jth column of X, $1 \leq j \leq n$. For each gene $1 \leq i \leq m$, consider the group-specific means $\bar{X}_{i,1} = n_1^{-1} \sum_{j=1}^{n_1} X_{ij}$ and $\bar{X}_{i,2} = n_2^{-1} \sum_{j=n_1+1}^{n} X_{ij}$. We assume that $\bar{X}_{i,1} \sim \mathscr{N}(\mu_{i,1}, \sigma_{i,1}^2/n_1)$ and $\bar{X}_{i,2} \sim \mathscr{N}(\mu_{i,2}, \sigma_{i,2}^2/n_2)$, where $\sigma_{i,1}^2$ and $\sigma_{i,2}^2$ denote the group-specific variances of individual observational units. The system of hypotheses of interest is given by $\mathscr{H}_m = (H_i : 1 \leq i \leq m)$, where $H_i = \{\mu_{i,1} = \mu_{i,2}\}, 1 \leq i \leq m$.

(i) Assuming $\sigma_{i,1}^2 = \sigma_{i,2}^2$, the statistic

$$T_i(X) = \sqrt{\frac{n_1 n_2}{n_1 + n_2}} \frac{\overline{X}_{i,1} - \overline{X}_{i,2}}{S}, \text{ where}$$

$$S^2 = \frac{1}{n-2} \left\{ \sum_{j=1}^{n_1} (X_{ij} - \overline{X}_{i,1})^2 + \sum_{j=n_1+1}^{n} (X_{ij} - \overline{X}_{i,2})^2 \right\},$$

is under H_i distributed as t_{n-2}, leading to the two-sided marginal p-value $p_i = 2(1 - F_{t_{n-2}}(|T_i(x)|))$, where x denotes the actually observed data.

(ii) If $\sigma_{i,1}^2 \neq \sigma_{i,2}^2$ has to be assumed, we consider the approximate version of the two-sample t-test proposed by Welch (1938). The underlying statistic is given by

$$T_i(X) = \frac{\overline{X}_{i,1} - \overline{X}_{i,2}}{\sqrt{\frac{S_{i,1}^2}{n_1} + \frac{S_{i,2}^2}{n_2}}}, \text{ where}$$

$$S_{i,1}^2 = \frac{1}{n_1 - 1} \sum_{j=1}^{n_1} (X_{ij} - \overline{X}_{i,1})^2,$$

$$S_{i,2}^2 = \frac{1}{n_2 - 1} \sum_{j=n_1+1}^{n} (X_{ij} - \overline{X}_{i,2})^2.$$

Under H_i, the distribution of $T_i(X)$ can be approximated by a t_ν-distribution, where the degrees of freedom ν are computed either by the method of Hsu (1938) or by the Welch–Satterthwaite formula, see Satterthwaite (1946) and Welch (1947). An approximate two-sided marginal p-value is thus given by $p_i = 2(1 - F_{t_\nu}(|T_i(x)|))$.

(b) Assume that the same n_1 individuals are measured under two different experimental conditions, such that $n = 2n_1$. Organize the data in the matrix X such that observational units j and n_1+j correspond to the two different conditions applied to individual j, where $1 \leq j \leq n_1$. For every gene $1 \leq i \leq m$, consider the row vector $D_i = (D_{i,1}, \ldots, D_{i,n_1})$ of differences, where $D_{ij} = X_{ij} - X_{i,n_1+j}$, $1 \leq j \leq n_1$. We assume that the mean $\bar{D}_i = n_1^{-1} \sum_{j=1}^{n_1} D_{ij}$ is normally distributed with mean μ_i and (in general unknown) variance σ_i^2/n_1. The system of hypotheses of interest is now given by $\mathcal{H}_m = (H_i : 1 \leq i \leq m)$, where $H_i = \{\mu_i = 0\}$, $1 \leq i \leq m$. Then, the statistic

$$T_i(X) = \sqrt{n_1}\frac{\bar{D}_i}{S_i}, \text{ where}$$

$$S_i^2 = \frac{1}{n_1 - 1}\sum_{j=1}^{n_1}(D_{ij} - \bar{D}_i)^2,$$

is under H_i distributed as t_{n_1-1}, leading to the two-sided marginal p-value $p_i = 2(1 - F_{t_{n_1-1}}(|T_i(x)|))$.

If the model additionally includes adjustments for covariates, one may assume that the Studentized regression coefficient corresponding to group membership is t-distributed, where the degrees of freedom are modified in the standard manner, and our considerations continue to apply. In any case, notice that Definition 10.1 does not define a complete statistical model, because only marginal distributions and p-values per gene were considered. At least for large sample sizes n, the assumption of a joint multivariate normal distribution for all m gene-specific mean expression differences seems justified. The following section is concerned with structural properties of the resulting covariance matrix for the two-sided test statistics.

10.2 Dependency Considerations

As discussed in Chaps. 4 and 5, the dependency structure among marginal test statistics or p-values, respectively, is crucial for choosing appropriate multiple test procedures. Lemma 10.1 shows that among the test statistics for differential gene expressions that we have discussed in Sect. 10.1, certain positive dependency relations hold true. For simplicity, we assume known marginal variances in Lemma 10.1.

Lemma 10.1 *Let Z_1 and Z_2 denote two standard normal random variables, where $Z = (Z_1, Z_2)^\top \sim \mathcal{N}_2(0, \Sigma)$, $\Sigma = \begin{pmatrix} 1 & \rho \\ \rho & 1 \end{pmatrix}$, $|\rho| < 1$. Then, the first two moments of $|Z| = (|Z_1|, |Z_2|)^\top$ are given by*

$$\mathbb{E}[|Z_1|] = \sqrt{2/\pi}, \; Var(|Z_1|) = 1 - 2/\pi, \tag{10.1}$$

$$\rho(|Z_1|, |Z_2|) = \frac{|\rho| \left\{ \pi - 2 \arctan\left(\frac{\sqrt{1-\rho^2}}{|\rho|} \right) \right\} + 2\left(\sqrt{1-\rho^2} - 1 \right)}{\pi - 2}. \tag{10.2}$$

Now, assume that the mean vector of Z_1 and Z_2, say $\mu = (\mu_1, \mu_2)^\top$, is unknown and that the two coordinate-wise hypotheses $H_1 = \{(\mu_1, \mu_2) \in \mathbb{R}^2 | \mu_1 = \mu_1^*\}$ and $H_2 = \{(\mu_1, \mu_2) \in \mathbb{R}^2 | \mu_2 = \mu_2^*\}$ are of interest. For a given local significance level $\alpha_{loc.}$, let $c = c(\alpha_{loc.}) = \Phi^{-1}(1 - \alpha_{loc.}/2)$ and consider the multiple test φ, given by $\varphi_j = \mathbf{1}_{(c,\infty)}(T_j)$ with $T_j = |Z_j - \mu_j^*|$, $j = 1, 2$. Then it holds

$$\mathbb{P}_{\mu^*}(T_2 \leq c \,|\, T_1 \leq c) = 1 - \frac{2}{1 - \alpha_{loc.}} \int_{-c}^{c} \phi(x) \Phi\left(\frac{\rho x - c}{\sqrt{1-\rho^2}} \right) dx. \tag{10.3}$$

Letting

$$\kappa(\rho) = \frac{\log\left(\mathbb{P}_{\mu^*}(T_2 \leq c \,|\, T_1 \leq c)\right)}{\log(1 - \alpha_{loc.})}, \tag{10.4}$$

we get

$$\alpha_{loc.} \leq FWER_{\mu^*}(\varphi) = 1 - (1 - \alpha_{loc.})^{1+\kappa(\rho)} \leq 1 - (1 - \alpha_{loc.})^2, \tag{10.5}$$

with equalities if and only if $Z_1 = Z_2$ almost surely, or Z_1 and Z_2 are stochastically independent, respectively.

Proof. The equations in (10.1) are well known and for instance reported at the end of Section 4 of Psarakis and Panaretos (2000). The right-hand side of (10.2) has been obtained by integrating the bivariate pdf of $|Z|$ as provided in Eq. (3.2) of Psarakis and Panaretos (2000) and is in line with the series expansion for $\rho(|Z_1|, |Z_2|)$ given in Appendix B of Asai and McAleer (2006). Equation (10.3) can be verified by elementary calculus, cf., e.g., Appendix A of Moskvina and Schmidt (2008). Finally, (10.5) can be seen by noticing that

$$\begin{aligned} \text{FWER}_{\mu^*}(\varphi) &= 1 - \mathbb{P}_{\mu^*}(T_2 \leq c, T_1 \leq c) \\ &= 1 - \mathbb{P}_{\mu^*}(T_1 \leq c)\mathbb{P}_{\mu^*}(T_2 \leq c \,|\, T_1 \leq c) \\ &= 1 - \exp\left\{\log(1 - \alpha_{\text{loc.}}) + \log(\mathbb{P}_{\mu^*}(T_2 \leq c \,|\, T_1 \leq c))\right\} \\ &= 1 - (1 - \alpha_{\text{loc.}})^{1+\kappa(\rho)}, \end{aligned}$$

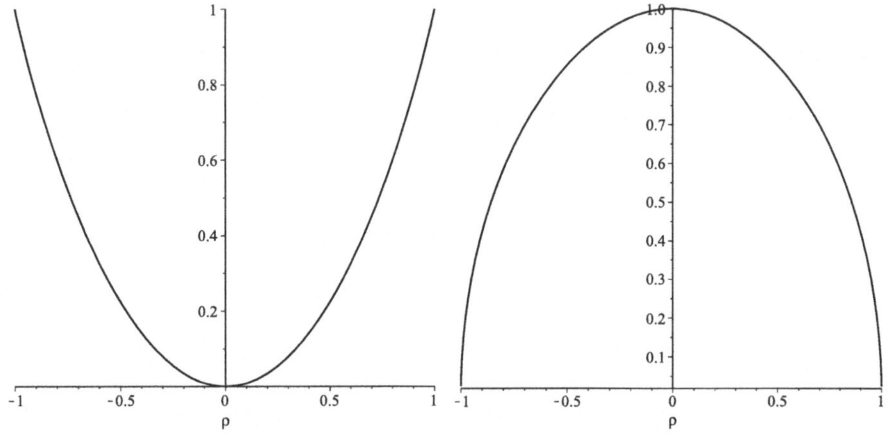

Fig. 10.1 Correlation coefficient $\rho(|Z_1|, |Z_2|)$ as in (10.2) (*left graph*) and $\kappa(\rho)$ as in (10.4) for $\alpha_{\text{loc.}} = 0.05$ (*right graph*), as functions of ρ

proving the equality relation in (10.5). Simple calculus yields that $\kappa(\rho)$ is decreasing in $|\rho|$, with maximum $\kappa(0) = 1$ and infimum $\lim_{|\rho| \to 1} \kappa(\rho) = 0$, completing the proof. □

Notice that $\rho(|Z_1|, |Z_2|)$ only depends on $|\rho|$ and is always non-negative, see the left graph in Fig. 10.1 for an illustration. The values of $\kappa(\rho)$ are depicted in the right graph in Fig. 10.1 for $\alpha_{\text{loc.}} = 0.05$.

Qualitatively, the positive dependency results reported in Lemma 10.1 remain to hold true if Studentization is performed in order to account for unknown variances. In the latter case, however, closed form expressions for the quantities corresponding to the ones in Lemma 10.1 are not so easily available. This is why we chose to present the results for the case of known variances for illustration.

In either case, the value of ρ is typically unknown in practice, but can be approximated by resampling, cf. our Sect. 3.2.1. For the particular case of applications in genetics, the book by Dudoit and van der Laan (2008) is a valuable reference. As explained in Chap. 8, the resampling methods of Dudoit and van der Laan (2008) are implemented in the `multtest` package for `R`. Iterating the reasoning of Lemma 10.1 results in a bound for the exceedance probability of an m-dimensional vector of test statistics for multiple testing of differential gene expression, where $m > 2$. Such a probability bound can be transformed into a simultaneous test procedure by virtue of our Sect. 4.3. However, this approach requires the estimation of an $m \times m$ covariance matrix which is a complicated task if m exceeds n. Alternatively to resampling, other techniques for covariance matrix estimation in high dimensions include shrinkage of the empirical covariance matrix $\hat{\Sigma}$ (say) toward some pre-specified target (cf., for instance, Schäfer and Strimmer (2005)) or low-rank approximations of $\hat{\Sigma}$ by, for example, assuming equi-correlation in blocks or AR(1) or Toeplitz structures, cf. Ghosal and Roy (2011) for applications in the context of multiple testing.

If one wants to avoid estimation of Σ, but still make use of the positive dependency results stated in Lemma 10.1, one can use stepwise rejective multiple tests (cf. Chap. 5) designed for applications with positively dependent test statistics or p-values, respectively. On the other hand, we have seen in Proposition 4.2 that conditions on the covariance matrix have to be imposed in order that absolute values of multivariate Gaussian vectors exhibit higher-order dependency relations like MTP_2 if the dimensionality m exceeds 2. Hence, a mathematically rigorous and comprehensive investigation of the behavior of stepwise rejective multiple tests which control the FWER or the FDR, respectively, under MTP_2 (cf. Table 5.1) is an open problem for absolute values of multivariate normal distributions. Related remarks can be found in the works of Block et al. (1993) and Glaz (2000). The case of absolute values of normal distributions in connection with the linear step-up test from Definition 5.6 has been treated by Reiner-Benaim (2007), see also the references therein. It turns out that a maximum FDR violation of $\alpha/8$ can occur, which is certainly negligible in practice for reasonable choices of the nominal FDR level α. Moreover, a positive result regarding multivariate t-distributions has recently been obtained by Block et al. (2013). Hence, we may conjecture that violations of the nominal FWER or FDR level, respectively, (if any) are of negligible magnitude in practice if the respective multiple tests from Table 5.1, which are guaranteed to work under MTP_2 or PRDS, are applied to vectors of test statistics for differential gene expression.

Furthermore, weak dependency in the sense of Definition 5.2 applies if the covariance matrix Σ has a block structure (as it is typically the case due to the genes' functional network structure); cf. Chap. 4 of Gontscharuk (2010). This entails asymptotic ($m \to \infty$) validity of the respective tests discussed in Chap. 5, at least if multiple power is asymptotically bounded away from zero.

10.3 Real Data Examples

10.3.1 Application of Generic Multiple Tests to Large-Scale Data

Notterman et al. (2001) published data from a cancer research project. The aim of the study was to identify differentially expressed gene and R(D)NA profiles in tumor tissue in comparison with normal (healthy) tissue. To this end, expression was assessed for 7457 different RNA, DNA and gene entities in 18 adenocarcinomic cancer patients. For each of these 18 study participants, the respective expression data were gathered once in cancer tissue and once in (paired) healthy tissue. The complete dataset is available as supplementary material to the article by Notterman et al. (2001).

After some Affymetrix preprocessing (cf. the "Materials and Methods" section in Notterman et al. (2001)), the comparison between the two paired tissue groups was performed by applying t-tests to the log-transformed data by utilizing the statistical model given in part (b) of Definition 10.1. This leads to $m = 7457$ marginal p-values.

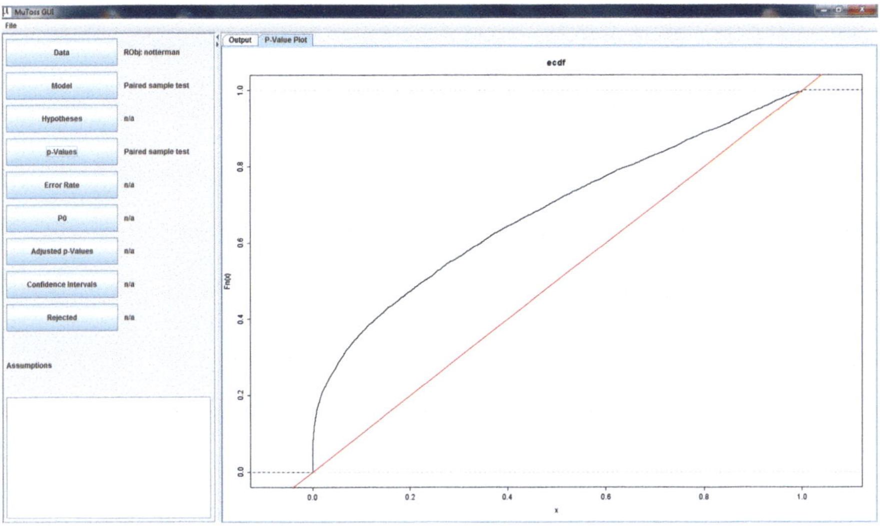

Fig. 10.2 Empirical cumulative distribution function of the $m = 7457$ marginal p-values resulting from the data by Notterman et al. (2001)

We re-analyzed the data with the μTOSS software, cf. Sect. 8.3. Figure 10.2 displays the ecdf of the $m = 7457$ marginal p-values. This plot is a helpful illustration with respect to stepwise rejective multiple tests, cf. the discussion around Lemma 5.6.

Assume that FWER control at level $\alpha = 5\,\%$ is targeted and no assumption about the (higher-order) dependency structure of the marginal p-values is imposed. Then, a generic multiple test keeping the chosen type I error criterion under any dependency structure is the Bonferroni–Holm test from Definition 5.3. Application of this step-down test to the marginal p-values results in $R_m(\varphi^{Holm}) = 113$ rejections, cf. Fig. 10.3.

The `mutoss` software implements the Bonferroni-Holm test by making use of the `mt.rawp2adjp` function from the `multtest` package, cf. Sect. 8.2. Thus, in addition to the mere rejection pattern, adjusted p-values for the rejected hypotheses are displayed in the rightmost column of the output list, see Fig. 10.3.

For comparison, applying the FDR criterion at level $\alpha = 5\,\%$ and the generically FDR-controlling step-up test φ^{BY} from Theorem 5.5 to this dataset, we obtain 305 additional rejections. However, we expect that among the $R_m(\varphi^{BY}) = 418$ rejections there are approximately 21 type I errors.

10.3.2 Copula Calibration for a Block of Correlated Genes

Let us consider a real-life dataset from cancer research which can be downloaded freely from the Gene Expression Omnibus data repository, see http://www.ncbi.nlm.nih.gov/sites/GDSbrowser, namely dataset GDS2771. Detailed information about

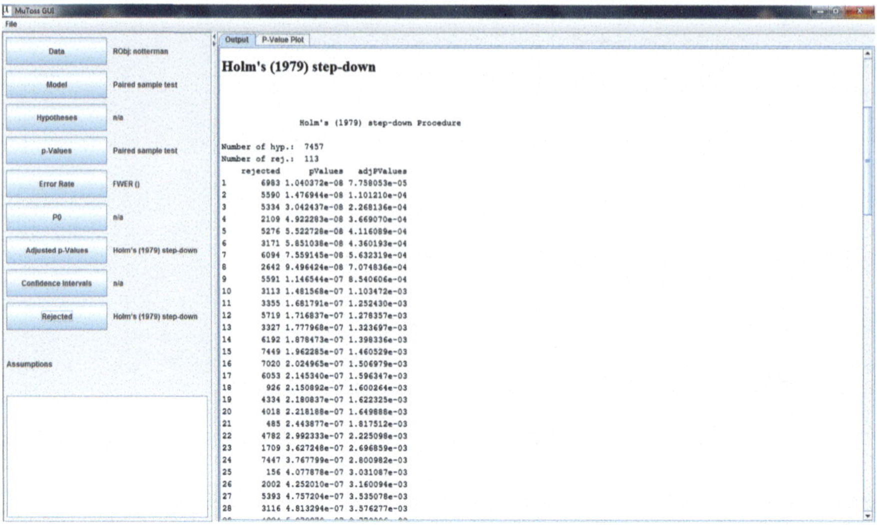

Fig. 10.3 Result of applying the Bonferroni–Holm test to the $m = 7457$ marginal p-values corresponding to the data by Notterman et al. (2001)

the underlying studies is given by Spira et al. (2007) and Gustafson et al. (2010). Dickhaus and Gierl (2013) focused on one specific aspect of this dataset, namely, the determination of genes that are (on average) differentially expressed in airway epithelial cells of cancer patients in comparison with healthy controls. To this end, they restricted their attention to $m = 11$ genes on chromosome 1, constituting a block of correlated genes. The considered genes have the identifiers `SYCP1`, `HS2ST1`, `RERE`, `PDE4DIP`, `CRP`, `SYT11`, `PAPPA2`, `PSEN2O`, `NENF`, `RAB3GAP2`, and `OBSCN`. In dataset GDS2771, expression profiles of $n_1 = 97$ patients and $n_2 = 90$ controls for these m genes are tabulated. In this, the raw expression counts were transformed in different ways in order to marginally fit normal distributions well. Indeed, diagnostic plots (not shown here) confirm that, marginally, Gaussian distributions are valid models. Consequently, the parameter of interest $\vartheta = (\vartheta_1, \ldots, \vartheta_m)^\top$ consists of the differences in mean expression levels of the $m = 11$ genes between the patient group and the control group on the corresponding transformed scales, with unique parameter value $\vartheta^* = \mathbf{0} \in \mathbb{R}^m$ in the global hypothesis H_0. Since all $n = 187$ observational units correspond to unrelated individuals, part (a) of Definition 10.1 has been applied. However, how the aforementioned (gene-specific) marginal transformations affect the dependency structure of p-values originating from marginal two-sample t-tests is not at all clear. Therefore, Dickhaus and Gierl (2013) chose to separate the dependency structure assessment completely from the marginal models (which is possible by the copula-based approach) and considered the flexible class of m-variate Clayton copulae (see, for instance, Example 4.23 in Nelsen (2006)) for the dependency modeling.

Each member of the family of Clayton copulae is uniquely defined by a one-dimensional parameter $\eta > 0$ and has the form

$$C_\eta(u_1, \ldots, u_m) = \left(u_1^{-\eta} + u_2^{-\eta} + \cdots + u_m^{-\eta} - m + 1\right)^{-1/\eta}. \qquad (10.6)$$

According to the copula-based method for constructing simultaneous test procedures that we have discussed in Sect. 4.4, Dickhaus and Gierl (2013) fitted an m-variate Clayton copula for approximating the dependency structure among the distributional transforms $1 - p_i$ of the t-statistics in each marginal $1 \leq i \leq m$. In this, they made use of the "Realized Copula" approach that we have described in Sect. 3.2.3. In order to assess the correlation structure of $(1 - p_i : 1 \leq i \leq m)$ under ϑ^*, a resampling strategy was employed. For a fixed number $B = 1{,}000$, the entire data vectors of the $n = n_1 + n_2 = 187$ study participants were permuted, i. e., randomly assigned to the "cancer positive" or the "cancer negative" group in each permutation run. This resampling mechanism destroys information about the differential expression between the two groups in every marginal (thus reflecting the situation under ϑ^*), but preserves the dependency structure between genes. After completion of all B permutations and re-calculation of the t-statistics in each permutation run, the empirical covariances of the resulting resampled distributional transforms were used as estimates $\hat{\sigma}_{ij}$ in the realized copula optimization step.

Based on this, application of the realized copula method to dataset GDS2771 with η taken as the Clayton copula parameter given in (10.6) resulted in $\hat{\eta} = 0.1636$, where each gene was treated equally, meaning that $\mathbf{W} = I_{\binom{11}{2}}$ was used. Having estimated the copula C_{ϑ^*} in this way, the reasoning of Theorem 4.8 led, for a target FWER level of $\alpha = 0.05$, to $\alpha_{\text{loc.}}^{(i)} \equiv \alpha_{\text{loc.}} = 0.00467$, $1 \leq i \leq m = 11$. Hence, the empirical calibration of $\alpha_{\text{loc.}}$ based on the intrinsic correlation structure in the data allowed for enlarging the multiplicity-adjusted local significance level in comparison with the Bonferroni correction or the Šidák correction.

10.4 LASSO and Statistical Learning Methods

High-dimensional data are nowadays often analyzed by regularized regression methods, in particular the LASSO (see our Sect. 7.3.1 for a brief introduction and the monograph by Bühlmann and van de Geer (2011) for a comprehensive theoretical treatment).

Wu et al. (2009) were concerned with LASSO penalized logistic regression in the context of genome-wide association analyses (cf. our Chap. 9, notice that both the endpoint and the predictors are categorical in this case). Wu (2005) proposed to fit a LASSO regression model for analyzing differential gene expression, but with the (quasi-)continuous expression profiles as endpoints and the dichotomous (or categorical) group indicators as covariates. This can be justified by considering that the direction of causation (if any) is unambiguous by the underlying biology. Furthermore, this approach is convenient from the point of view of implementation. Anyhow,

Garcia-Magariños et al. (2010) developed LASSO logistic regression algorithms for analyzing differential gene expression, where the phenotype is considered as the response and the expression profiles as the vector of covariates. This may be considered as the most appropriate LASSO-based data analysis strategy for differential gene expression.

The methods discussed in this section have a high potential for classification and diagnostic purposes, because a full multivariate model for the class-specific gene expression profiles is learned. This also holds true for statistical (machine) learning methods like support vector machines (SVMs), see Part II in the book of Vapnik (1998) for a theoretical introduction from the statistical learning perspective, Blanchard et al. (2008) for a theoretical treatment from the statistical point of view and Brown et al. (2000) for an application to gene expression data, among many others. At present, the drawback of such regularized regression methods and statistical learning models is that inferential theory for the resulting effect size estimates is not well-developed yet, such that their usage for the purpose of assessing statistical significance of differential expression profiles is much less straightforward than for the methods discussed in Sect. 10.3, cf. our Sect. 7.3.1 for some recent results about inferential methods based on the LASSO. Hence, statistical methodology has to be chosen according to the actual aim of the analysis.

On the other hand, there are even problems where the two inferential problems multiple testing and classification occur at the same time. One particular such use case occurs in biomarker studies if the aim is to classify subjects into disease groups on the basis of their biomarker profiles. Typically, in a first stage a subset of relevant markers has to be selected from the very large set of all available biomarkers (a multiple testing problem). Then, in a second stage, classification of subjects is performed on the basis of feature vectors built from the selected biomarkers. Such a two-stage design is chosen by Freidlin et al. (2010), for example. They are considered with the specific problem of identifying a subgroup of cancer patients which is responsive to treatment on the basis of gene expression levels. Freidlin et al. (2010) employ a resampling scheme for the entire two-stage procedure in order to assess statistical significance of treatment effects in the identified subgroup. This can be regarded as a proxy for classification accuracy in this particular context.

10.5 Gene Set Analyses and Group Structures

Since genes are not isolated biological units, but organized in groups and networks, one strand of modern research in statistical genetics is concerned with the problem of integrating experimental genomic data and exogenous functional information as provided by gene ontology terms, for example. From the point of view of statistical modeling, this leads to inference for graph- or tree-structured data.

Newton et al. (2012) consider cases in which the experimental data are measured at the level of genes, but inference is required at the level of functional categories. They propose a probabilistic graphical modeling approach for such functional-category inference under the Bayesian paradigm. In particular, their graphical "role model" is capable of reflecting that genes can have different functions depending on the biological context. See also the references in Newton et al. (2012) for earlier developments in this direction.

Frequentist methods for gene set analyses are often based (sometimes implicitly) on the closed test principle, cf. our discussion at the end of Sect. 3.3. For specific applications in gene expression analyses see, for instance, Goeman et al. (2004), Mansmann and Meister (2005), Goeman and Bühlmann (2007), Hummel et al. (2008), Goeman and Mansmann (2008), and Goeman and Finos (2012). The general idea underlying all these methods is that the graph- or tree-structure imposes logical constraints on the system of hypotheses, such that not all combinations of true/false hypotheses which are possible in general can actually occur, because some of these combinations contradict the relations expressed by edges in the gene network graph. Computationally feasible solutions are worked out to test the remaining hypotheses in the induced system of intersection hypotheses in an efficient manner.

Acknowledgments Data analysis in Sect. 10.3.2 is joint work with Jakob Gierl.

References

Asai M, McAleer M (2006) Asymmetric multivariate stochastic volatility. Econ Rev 25(2–3):453–473. doi:10.1080/07474930600712913

Blanchard G, Bousquet O, Massart P (2008) Statistical performance of support vector machines. Ann Stat 36(2):489–531. doi:10.1214/009053607000000839

Block HW, Costigan TM, Sampson AR (1993) Optimal second-order product probability bounds. J Appl Probab 30(3):675–691. doi:10.2307/3214774

Block HW, Savits TH, Wang J, Sarkar SK (2013) The multivariate-*t* distribution and the Simes inequality. Stat Probab Lett 83(1):227–232. doi:10.1016/j.spl.2012.08.013

Brown MPS, Noble Grundy W, Walsh Sugnet C (2000) Knowledge-based analysis of microarray gene expression data by using support vector machines. Proc Natl Acad Sci USA 97(1):262–267

Bühlmann P, van de Geer S (2011) Statistics for high-dimensional data: Methods, theory and applications. Springer series in statistics. Springer, Berlin. doi: 10.1007/978-3-642-20192-9

Dickhaus T, Gierl J (2013) Simultaneous test procedures in terms of p-value copulae. Global Science and Technology Forum (GSTF). In: Proceedings on the 2nd annual international conference on computational mathematics, computational geometry and statistics (CMCGS 2013), vol 2, pp 75–80

Dudoit S, van der Laan MJ (2008) Multiple testing procedures with applications to genomics. Springer series in statistics. Springer, New York

Freidlin B, Jiang W, Simon R (2010) The cross-validated adaptive signature design. Clin Cancer Res 16(2):691–698

Garcia-Magariños M, Antoniadis A, Cao R, Gonzalez-Manteiga W (2010) LASSO logistic regression, GSoft and the cyclic coordinate descent algorithm: application to gene expression data. Stat Appl Genet Mol Biol 9:Article30.

Ghosal S, Roy A (2011) Predicting false discovery proportion under dependence. J Am Stat Assoc 106(495):1208–1218. doi:10.1198/jasa.2011.tm10488

Glaz J (2000) Probability inequalities for multivariate distributions with applications to statistics. Chapman and Hall/CRC Press, Boca Raton

Goeman JJ, Bühlmann P (2007) Analyzing gene expression data in terms of gene sets: methodological issues. Bioinformatics 23(8):980–987

Goeman JJ, Finos L (2012) The inheritance procedure: multiple testing of tree-structured hypotheses. Stat Appl Genet Mol Biol 11(1):Article 11

Goeman JJ, Mansmann U (2008) Multiple testing on the directed acyclic graph of gene ontology. Bioinformatics 24(4):537–544

Goeman JJ, van de Geer SA, de Kort F, van Houwelingen HC (2004) A global test for groups of genes: testing association with a clinical outcome. Bioinformatics 20(1):93–99

Gontscharuk V (2010) Asymptotic and exact results on FWER and FDR in multiple hypotheses testing. Ph.D. thesis, Heinrich-Heine-Universität Düsseldorf

Gustafson AM, Soldi R, Anderlind C, Scholand MB, Qian J, Zhang X, Cooper K, Walker D, McWilliams A, Liu G, Szabo E, Brody J, Massion PP, Lenburg ME, Lam S, Bild AH, Spira A (2010) Airway PI3K pathway activation is an early and reversible event in lung cancer development. Sci Transl Med 2(26):26ra25

Hsu P (1938) Contribution to the theory of "student's" t-test as applied to the problem of two samples. Statist Res Mem, Univ London 2:1–24

Hummel M, Meister R, Mansmann U (2008) GlobalANCOVA: exploration and assessment of gene group effects. Bioinformatics 24(1):78–85

Mansmann U, Meister R (2005) Testing differential gene expression in functional groups. Goeman's global test versus an ANCOVA approach. Methods Inf Med 44(3):449–453

Moskvina V, Schmidt KM (2008) On multiple-testing correction in genome-wide association studies. Genet Epidemiol 32:567–573

Nelsen RB (2006) An introduction to copulas. Springer series in statistics, 2nd edn. Springer, New York

Newton MA, He Q, Kendziorski C (2012) A model-based analysis to infer the functional content of a gene list. Stat Appl Genet Mol Biol 11(2):Article 9

Notterman DA, Alon U, Sierk AJ (2001) Transcriptional gene expression profiles of colorectal adenoma, adenocarcinoma, and normal tissue examined by oligonucleotide arrays. Cancer Res 61:3124–3130

Psarakis S, Panaretos J (2000) On some bivariate extensions of the folded normal and the folded t distributions. J Appl Stat Sci 10(2):119–136

Reiner-Benaim A (2007) FDR control by the BH procedure for two-sided correlated tests with implications to gene expression data analysis. Biom J 49(1):107–126

Satterthwaite FE (1946) An approximate distribution of estimates of variance components. Biometrics Bull 2(6):110–114. doi:10.1214/aoms/1177730491

Schäfer J, Strimmer K (2005) A shrinkage approach to large-scale covariance matrix estimation and implications for functional genomics. Stat Appl Genet Mol Biol 4:Article32

Smyth GK (2004), Linear models and empirical Bayes methods for assessing differential expression in microarray experiments. Stat Appl Genet Mol Biol 3:Article3

Spira A, Beane JE, Shah V, Steiling K, Liu G, Schembri F, Gilman S, Dumas YM, Calner P, Sebastiani P, Sridhar S, Beamis J, Lamb C, Anderson T, Gerry N, Keane J, Lenburg ME, Brody JS (2007) Airway epithelial gene expression in the diagnostic evaluation of smokers with suspect lung cancer. Nat Med 13(3):361–366

Tusher VG, Tibshirani R, Chu G (2001) Significance analysis of microarrays applied to the ionizing radiation response. Proc Natl Acad Sci USA 98(9):5116–5121

Vapnik VN (1998) Statistical learning theory. Adaptive and learning systems for signal processing, communications, and control. Wiley, Chichester

Welch B (1947) The generalization of student's problem when several different population variances are involved. Biometrika 34:28–35

Welch BL (1938) The significance of the difference between two means when the population variances are unequal. Biometrika 29:350–362

Wu B (2005) Differential gene expression detection using penalized linear regression models: the improved SAM statistics. Bioinformatics 21(8):1565–1571

Wu TT, Chen YF, Hastie T, Sobel E, Lange K (2009) Genome-wide association analysis by LASSO penalized logistic regression. Bioinformatics 25(6):714–721

Chapter 11
Functional Magnetic Resonance Imaging

Abstract Analyzing functional magnetic resonance imaging (fMRI) data is a particular challenge for statistical methodology and practice, because such data are high-dimensional and spatially and temporally correlated. Furthermore, the questions of scientific interest often relate to so-called regions of interest constituted by clusters in the sample space. We are first considered with spatial models and describe false discovery rate-controlling multiple tests for continuous families of hypotheses and for grouped systems of hypotheses. Then, we show how random field theory can be utilized to construct family-wise error rate-controlling simultaneous test procedures which account for the topological structure of the brain. Finally, multivariate time series models are presented which are capable of modeling both the spatial and the dynamic components of the signals in the data.

If a genetic association or expression study has detected associations between the phenotype of interest and genes which are related to brain activity, a functional magnetic resonance imaging (fMRI) experiment can be carried out in order to confirm the association functionally. This reduces the biological distance between genotype and phenotype.

Functional magnetic resonance imaging is an indirect (non-invasive) way to measure brain activity. It consists in imaging the change in blood flow (hemodynamic response) related to energy use by brain cells. The blood oxygen level dependent (BOLD) response is thereby measured in three-dimensional volume units (voxels) over a certain period of time by a series of time-discrete scans (images). The technical devices used to carry out these measurements are called fMRI scanners. The data structure arising from an fMRI experiment of one particular individual is a four-dimensional array, where the first (say) three dimensions correspond to the spatial location in the brain and the last dimension indexes time on a discrete time axis corresponding to the scans. The challenge for statistical methodology is that such data are high-dimensional and spatially and temporally correlated. However, the questions of scientific interest often do not relate to the voxels as observational units themselves, but to "regions of interest" (ROIs) constituted by clusters of voxels. Such ROIs may be explicitly pre-defined according to prior knowledge about the anatomy

T. Dickhaus, *Simultaneous Statistical Inference*,
DOI: 10.1007/978-3-642-45182-9_11,

or the function of the brain, or data-adaptively by a cluster analysis in a pre-study (cf. Poldrack (2007) and Heller et al. (2006) for more details). In either case, the cluster structure induces a hierarchy (and a grouping) in the voxel space and can be used to reduce dimensionality.

11.1 Spatial Modeling

The spatial components of the BOLD responses of n observational units can be modeled as random fields $\{Y_i(s) : s \in \mathscr{S}\}$, $1 \leq i \leq n$, where $\mathscr{S} \subset \mathbb{R}^3$ denotes a manifold describing the brain or the brain region of interest. Often, the spatial structure of the BOLD response is the target of statistical inference and hence, the voxel-wise autocorrelation structure can be regarded as nuisance. If this is the case, typically whitening is performed to every voxel-wise time series in order to obtain a valid summary of the brain activity at the respective spatial location; see, e.g., Worsley et al. (2002) and references therein for autoregressive modeling of the autocorrelation structure in fMRI voxels. Typically, also spatial smoothing is applied to the raw data to account for the discrete grid structure of the voxels in the scans. For a detailed discussion about the design of fMRI experiments and different techniques for data preprocessing we defer the reader to Lazar (2008).

After the aforementioned and possibly further pre-processing steps, we follow the modeling approach by Worsley (2003) and Taylor and Worsley (2007). We assume a linear model of the form

$$Y_i(s) = \mathbf{x}_i \vartheta(s) + \sigma(s)\varepsilon_i(s), \tag{11.1}$$

where $\mathbf{x}_i$ denotes a k-dimensional row vector of known regressors (covariates) of observational unit i, $\vartheta(s)$ a location-specific, unknown vector of k regression coefficients, and $(\varepsilon_i(s) : 1 \leq i \leq n)$ are stochastically independent Gaussian random fields, each with mean zero and unit variance. The location-specific variance $\sigma(s)$ is typically unknown and estimated from the repetitions (where whitening plays an important role if these are autocorrelated repetitions over time). In the case of mere (spatial) signal detection problems, k may be equal to one and $x_i = 1$ for all $1 \leq i \leq n$. Observational units may relate to different subjects (where a standardization with respect to brain topology constitutes a further pre-processing step) or to repetitions of stimulus presentation to the same subject, among other paradigms.

In any case, we assume that inference is targeted at some pre-defined contrasts regarding $\vartheta(s)$, $s \in \mathscr{S}$. This leads to a system $(H_s : s \in \mathscr{S})$ of hypotheses in the spatial domain. Furthermore, we assume that a method (typically some kind of t-test) exists to transform the original data into a Gaussian-related random field $(T(s) : s \in \mathscr{S})$ of test statistics for the respective spatial positions, where the distribution of $T(s)$ under H_s is known such that $T(s)$ can be transformed into a valid p-value for testing H_s.

For the case that $\mathscr{S}$ is the set of voxel locations itself, application of multiple testing methods that we have discussed in previous chapters is rather straightforward. For example, Genovese et al. (2002) proposed to control the FDR on the basis of individual voxels. They argued that the linear step-up test from Definition 5.6 is appropriate, because presence of only non-negative correlations among the Gaussian error terms per voxel "may be a reasonable assumption for many fMRI data sets". For the more general case that $\mathscr{S}$ may be a spatial continuum, the following generalization of the FDR has been proposed by Pacifico et al. (2004) and applied by Benjamini and Heller (2007) and Blanchard et al. (2013), among others.

Definition 11.1 (Pacifico et al. (2004)). Assume that $\mathscr{S}$, equipped with some σ-field, is a measurable space and that $\lambda : \mathscr{S} \to \mathbb{R}_{\geq 0}$ is some measure on $\mathscr{S}$. Let φ be a multiple test procedure for the system $(H_s : s \in \mathscr{S})$ of hypotheses and $A = A(\varphi)$ the random set of hypotheses that are rejected by φ. Denote by $\mathscr{S}_0$ the subset of $\mathscr{S}$ on which the H_s hold true. Then, the false discovery rate of φ with respect to λ is defined by

$$\mathrm{FDR}_{\lambda}(\varphi) = \mathbb{E}_{\vartheta}\left[\frac{\lambda(A \cap \mathscr{S}_0)}{\lambda(A)} \mathbf{1}_{\{\lambda(A)>0\}}\right], \qquad (11.2)$$

where ϑ denotes the parameter of the underlying statistical model.

As discussed by Blanchard et al. (2013), testing a continuum of null hypotheses may induce measurability issues in general. Hence, we assume here that all objects of interest are well-defined. Blanchard et al. (2013) also introduce a generalized class of step-up procedures which offer control of FDR_{λ}. These procedures are continuous-space analogues of the linear step-up test φ^{LSU} from Definition 5.6 (applicable under a continuous-space analogue of the PRDS condition regarding the joint distribution of p-values) and the family φ^{ν} of tests discussed in Theorem 5.6 (applicable without any assumption on the dependency structure).

11.2 False Discovery Rate Control for Grouped Hypotheses

11.2.1 Clusters of Voxels

Returning to the ROI approach and following Benjamini and Heller (2007), let us consider a finite partition of $\mathscr{S}$ into m contiguous components $C_1, \ldots, C_m$, called clusters. Then, the finite system $\mathscr{H}_m = (H_i : 1 \leq i \leq m)$ of null hypotheses with corresponding alternatives K_i, $1 \leq i \leq m$, is of interest, where

$$H_i = \bigcap_{s \in C_i} H_s, \text{ versus } K_i = \bigcup_{s \in C_i} K_s. \qquad (11.3)$$

In (11.3), H_s and K_s refer to the hypotheses at the individual spatial locations as discussed before. The interpretation of testing the pair of hypotheses defined in (11.3) is that a cluster C_i is considered "active" as soon as at least one H_s with $s \in C_i$ is false. Certainly, clusters are in general heterogeneous, for instance with respect to their λ-cardinality, where λ is as in Definition 11.1. In order to incorporate this cluster heterogeneity into the statistical methodology, a weighted version of the FDR criterion, originally introduced by Benjamini and Hochberg (1997), seems appropriate.

Definition 11.2 (cf. Benjamini and Hochberg (1997)). Assume that m non-negative weights w_i, $1 \leq i \leq m$, are given such that $\sum_{i=1}^{m} w_i = m$. Let $I_0 \subseteq \{1, \ldots, m\}$ denote the subset of clusters fulfilling that H_i is true for all $i \in I_0$, where H_i is defined in (11.3), and let $\varphi = (\varphi_i : 1 \leq i \leq m)$ denote a multiple test for the system $\mathscr{H}_m = (H_i : 1 \leq i \leq m)$ of cluster hypotheses. Consider random binary indicators $A_1 = A_1(\varphi), \ldots, A_m = A_m(\varphi)$ such that $A_i = 1$ if and only if hypothesis H_i is rejected by φ. Then, the weighted FDR of φ under ϑ is given by

$$\text{wFDR}_\vartheta(\varphi) = \mathbb{E}_\vartheta \left[\frac{\sum_{i \in I_0} w_i A_i}{\sum_{i=1}^{m} w_i A_i} \mathbf{1}_{\{\sum_{i=1}^{m} A_i > 0\}} \right]. \tag{11.4}$$

Notice that $\text{wFDR}_\vartheta(\varphi) = \text{FDR}_\vartheta(\varphi)$ in the case that uniform weights $w_1 = \cdots = w_m = 1$ are used. Benjamini and Heller (2007) discuss two wFDR-controlling multiple tests, namely, a weighted version of φ^{LSU} from Definition 5.6 and a data-adaptive two-stage step-up test similar to the ones discussed by Benjamini et al. (2006) (see Sect. 3.1.3).

After the application of a multiple test on cluster basis, it may be of interest to try to localize the false location-specific hypotheses H_s within each rejected cluster. To this end, Benjamini and Heller (2007) define conditional within-cluster p-values which are valid for testing the individual hypotheses H_s, $s \in C_i$, given that cluster C_i has been rejected. Since these p-values are valid conditionally to the event that cluster C_i has been rejected in the first stage of the statistical analysis, they may be used in standard FDR-controlling tests, cf. Chap. 5.

A different weighting approach has been pursued by Hu et al. (2010). The authors are considered with control of the FDR on the basis of a finite system of individual hypotheses (which may in the fMRI context relate to voxels), but propose to utilize the group structure among these hypotheses for a weighting of the individual p-values. In particular, weights based on the estimated proportion of true hypotheses in each group separately are discussed. Further results regarding group-wise testing and weighting of hypotheses under the FDR paradigm have been derived by Bogomolov (2011).

It may be remarked here that routines for cluster-based FDR-controlling multiple tests with weights are not yet implemented in the μTOSS software (cf. Sect. 8.3), but are available from the original authors; see, for instance, http://www.math.tau.ac.il/~ybenja/software.html.

11.2.2 Multiple Endpoints per Location

In some applications of fMRI, brain activity is analyzed with respect to several endpoints (for instance, cognitive tasks) simultaneously. This induces a second layer of multiplicity and leads to multivariate test statistics or p-values, respectively. Benjamini and Heller (2008) distinguish between several types of scientific questions in this context, which can be formalized by different families of hypotheses.

Definition 11.3 (Benjamini and Heller (2008)). Consider a finite set of S spatial locations and m different endpoints that are to be analyzed at every spatial location $1 \leq s \leq S$. This leads to a system of $m \times S$ individual hypotheses, given by $(H_\ell(s) : 1 \leq \ell \leq m,\ 1 \leq s \leq S)$. Different possibilities for aggregating the m hypotheses per spatial locations s are given by the following families of hypotheses.

$$H^{\text{conj.}}(s) = \bigcap_{1 \leq \ell \leq m} H_\ell(s) \quad \text{versus} \quad K^{\text{conj.}}(s) = \bigcup_{1 \leq \ell \leq m} K_\ell(s), \tag{11.5}$$

$$H^{\text{disj.}}(s) = \bigcup_{1 \leq \ell \leq m} H_\ell(s) \quad \text{versus} \quad K^{\text{disj.}}(s) = \bigcap_{1 \leq \ell \leq m} K_\ell(s), \tag{11.6}$$

$$H^{u/m}(s) = m_1(s) < u \quad \text{versus} \quad K^{u/m}(s) = m_1(s) \geq u, \ \ 1 \leq u \leq m, \tag{11.7}$$

where $m_1(s) = \#\{1 \leq \ell \leq m : H_\ell(s) \text{ is false}\}$. Notice that $H^{1/m}(s) = H^{\text{conj.}}(s)$ and $H^{m/m}(s) = H^{\text{disj.}}(s)$. Hypothesis $H^{\text{conj.}}(s)$ is called the conjunction hypothesis, $H^{u/m}(s)$ for $1 < u < m$ a partial conjunction hypothesis, and $H^{\text{disj.}}(s)$ the disjunction hypothesis at location s, respectively.

Noticing the similarity between testing $H^{\text{conj.}}(s)$ and FWER control in the weak sense (cf. part (g) of Definition 1.2) it is fair to argue that testing $H^{\text{conj.}}(s)$ is a too weak criterion. On the other hand, it will typically hardly be possible to reject $H^{\text{disj.}}(s)$ if m is moderate or large, such that testing the "intermediate" hypotheses $H^{u/m}(s)$ for $2 \leq u \leq m-1$ may be considered a suitable compromise, provided that $m > 2$. Motivated by the Simes test (cf. Sect. 5.3), Benjamini and Heller (2008) proved the following theorem about valid p-values for testing $(H^{u/m}(s) : 1 \leq s \leq S)$.

Theorem 11.1 (Benjamini and Heller (2008)). *For every spatial location $1 \leq s \leq S$, assume that valid p-values $p_1(s), \ldots, p_m(s)$ for testing $(H_\ell(s) : 1 \leq \ell \leq m)$ are available and denote their order statistics by $p_{1:m}(s) \leq \cdots \leq p_{m:m}(s)$. For given parameter u as in Definition 11.3, let*

$$p^{u/m}(s) = \min_{1 \leq \ell \leq m-u+1} \left\{ \frac{m-u+1}{\ell} p_{u-1+\ell:m}(s) \right\}.$$

If the joint distribution of $p_1(s), \ldots, p_m(s)$ fulfills the PRDS condition, then $p^{u/m}(s)$ is a valid p-value for testing $H^{u/m}(s)$.

Of course, if the PRDS assumption in Theorem 11.1 can not be established, a Bonferroni-type adjustment can be employed, leading to $p^{u/m}(s) = (m-u+1)p_{u:m}(s)$. In practice, it remains to choose u, which may introduce some arbitrariness in the statistical analysis. Benjamini and Heller (2008) propose to test $H^{u/m}(s)$ for all $1 \leq u \leq m$ and to superimpose the m rejection patterns in a graphical display. Furthermore, they propose a "second-layer" adjustment for multiplicity with respect to these m tests per location.

The usage of partial conjunction hypotheses in connection with FWER control has been discussed by Friston et al. (2005). However, due to the massive multiplicity given by S and m, FDR control appears to be the more appropriate criterion, at least for screening purposes.

11.3 Exploiting Topological Structure by Random Field Theory

Adopting the notation of Definition 11.1, the continuous-space analogue of the FWER of a multiple test φ with respect to the measure λ is given by

$$\text{FWER}_{\lambda}(\varphi) = \mathbb{P}_{\vartheta}(\lambda(A(\varphi) \cap \mathcal{S}_0) > 0) \tag{11.8}$$

and a multiplicity-adjusted p-value at location s corresponding to a simultaneous test procedure (see Chap. 4) is given by

$$\mathbb{P}_0 \left(\max_{\tilde{s} \in \mathcal{S}} T(\tilde{s}) \geq t(s) \right), \tag{11.9}$$

where $t(s)$ denotes the actually observed value of T at location s and $\mathbb{P}_0$ the measure under the global hypothesis that all H_s are true. For ease of argumentation, we assume that $\mathbb{P}_0$ is uniquely defined, an assumption that typically holds true in relevant applications. Equation (11.9) relates control of FWER_{λ} by means of simultaneous test procedures directly to the Euler characteristic heuristic from (4.20) in Sect. 4.5.

Since the human brain has a complex topological structure and the spatial structure of the noise covariances will in general not be trivial, a nonisotropic field $(T(s) : s \in \mathcal{S})$ has to be considered and Lipschitz-Killing curvatures $\mathcal{L}_j(\mathcal{S})$ have to be estimated for $0 \leq j \leq 3$. To this end, the estimation approach by Worsley et al. (1999), Worsley (2003) and Taylor and Worsley (2007) is convenient. First, assume for a moment that the field would be isotropic. Then, the $\mathcal{L}_j(\mathcal{S})$ in formula (4.21) could be replaced by the intrinsic volumes of $\mathcal{S}$, say $\mu_j(\mathcal{S})$, $0 \leq j \leq 3$. The numbers $\mu_j(\mathcal{S})$ only depend on the geometry of $\mathcal{S}$, where exact expressions are provided as formula (4) by Worsley (2003) and formula (9) by Taylor and Worsley (2007), respectively. In the general (nonisotropic) case, consider the (normalized) least squares residuals under model (11.1), given by

$$r(s) = Y(s) - X(X^\top X)^{-1} X^\top Y(s),$$
$$u(s) = r(s)/||r(s)||,$$

where $Y(s) = (Y_1(s), \ldots, Y_n(s))^\top$ and X is the design matrix of model (11.1), the i-th row of which is equal to $\mathbf{x}_i$, $1 \le i \le n$. Notice that, for any $1 \le i \le n$, $(r_i(s) : s \in \mathscr{S})$ is a mean-zero Gaussian random field with the same spatial correlation structure as the corresponding noise component $(\varepsilon_i(s) : s \in \mathscr{S})$ of model (11.1). Based on this relationship, the proposed estimators are given by

$$\widehat{\mathscr{L}_j}(\mathscr{S}) = \mu_j(u(\mathscr{S})), \ 0 \le j \le 3.$$

Software for fMRI data analysis based on random field theory is available, cf. the survey in Appendix A of Lazar (2008).

11.4 Spatio-Temporal Models via Multivariate Time Series

In the previous sections, we have focused our attention on the spatial structure of the BOLD response and have regarded the autocorrelation structure as nuisance. Here, we study multivariate time series models which explicitly take into account this autocorrelation structure. For a particular voxel i, this leads to a time series, say $(X_i(t) : 1 \le t \le T)$, where T denotes the number of scans or time points. Consequently, the complete fMRI dataset corresponding to one individual is a multivariate time series $(\mathbf{X}(t) : 1 \le t \le T)$, where $\mathbf{X}(t) = (X_1(t), \ldots, X_p(t))^\top$ and p denotes the number of spatial locations (voxels).

Modeling of the BOLD response as a multivariate time series offers the possibility to analyze both spatial and temporal structures. However, due to the high dimensionality p, standard statistics to analyze the dynamic behavior of times series, like for instance the empirical autocovariance function, are severely ill-conditioned in our setting. Hence, some kind of dimension reduction or regularization of the time series model is needed in order to perform statistical inference reliably. One solution is to consider a factor model. The underlying assumption is that the dynamic behavior of the process $\mathbf{X}$ can already be described well (or completely) by a lower-dimensional, possibly latent process. This leads to the following type of model.

Definition 11.4. The multivariate time series model for the observable process $\mathbf{X}$ is called a dynamic factor model (DFM), if

$$\mathbf{X}(t) = \sum_{s=-\infty}^{\infty} \Lambda(s)\,\mathbf{f}(t-s) + \varepsilon(t), \ 1 \le t \le T. \qquad (11.10)$$

In (11.10), $\mathbf{X} = (\mathbf{X}(t) : 1 \le t \le T)$ denotes a p-dimensional, covariance-stationary stochastic process in discrete time with mean zero, $\mathbf{f}(t) = (f_1(t), \ldots, f_k(t))^\top$ with $k < p$ denotes a k-dimensional vector of so-called "common factors" and

$\varepsilon(t) = (\varepsilon_1(t), \ldots, \varepsilon_p(t))^\top$ denotes a p-dimensional vector of "specific factors", to be regarded as error or remainder terms. Both $\mathbf{f}(t)$ and $\varepsilon(t)$ are assumed to be centered and the error terms are modeled as noise in the sense that they are mutually uncorrelated at every time point and, in addition, uncorrelated with $\mathbf{f}(t)$ at all leads and lags. The entry (i, j) of the matrix $\Lambda(s)$ is called a "factor loading" and quantitatively reflects the influence of the j-th common factor at lead or lag s, respectively, on the i-th component of $\mathbf{X}(t)$, where $1 \le i \le p$ and $1 \le j \le k$.

A special case of model (11.10) results if the same factor loading matrix Λ (say) is assumed at all leads and lags, such that

$$\mathbf{X}(t) = \Lambda \mathbf{Z}(t) + \varepsilon(t), \; 1 \le t \le T, \tag{11.11}$$

with common factors $\mathbf{Z}(t) = (Z_1(t), \ldots, Z_k(t))^\top$ (say). Peña and Box (1987) were concerned with methods for the determination of the (number of) common factors in a factor model of the form (11.11) and derived a canonical transformation allowing a parsimonious representation of $\mathbf{X}(t)$ in (11.11) in terms of the common factors. As noted by Park et al. (2009), the model (11.10) can be accommodated into (11.11) if the common factors are themselves considered as unknown model parameters. This can be seen by considering each Z_j, $1 \le j \le k$, as a lagged linear combination of the common factors $\mathbf{f}$ in (11.10). Further relationships between different types of dynamic factor models are explained by Hallin and Lippi (2013), for example; see also the references therein.

Under model (11.11), the univariate time series model corresponding to voxel $1 \le i \le p$ is given by

$$\begin{aligned} X_i(t) &= \sum_{j=1}^{k} \Lambda_{i,j} Z_j(t) + \varepsilon_i(t) \\ &= \Lambda_{i,1} Z_1(t) + \Lambda_{i,2} Z_2(t) + \cdots + \Lambda_{i,k} Z_k(t) + \varepsilon_i(t). \end{aligned}$$

Hence, by symmetry, one may also interpret the columns of Λ as spatial factors and the values $(Z_1(t), \ldots, Z_k(t))^\top$ as time-dependent factor loadings. An interesting idea was advocated in Park et al. (2009) and van Bömmel et al. (2013). The authors propose to model the spatial factors as functions of the spatial locations, such that $\Lambda_{i,j} = \lambda_j(i_1, i_2, i_3)$, where (i_1, i_2, i_3) encodes the spatial position of voxel i. Estimating the functions $\lambda_1, \ldots, \lambda_k$ nonparametrically by a low-dimensional space basis reduces model complexity further such that inference becomes feasible, even in the very high-dimensional voxel space. The estimated functions $\hat{\lambda}_1, \ldots, \hat{\lambda}_k$ approximate the spatial structure of the BOLD response, while the estimated time-dependent factor loadings $\hat{Z}_1(t), \ldots, \hat{Z}_k(t)$, $1 \le t \le T$, capture the dynamic structure.

If small regions of interest are studied, inference can even be targeted at a full parametric representation of the DFM given in (11.10). To this end, it is convenient to study the frequency-domain representation of the process. The model equation (11.10) immediately entails that the autocovariance function of the observable process $\mathbf{X}$, $\Gamma_{\mathbf{X}}$ for short, and its spectral density matrix $S_{\mathbf{X}}$ (say), can be expressed by

$$\Gamma_{\mathbf{X}}(u) = \mathbb{E}[\mathbf{X}(t)\mathbf{X}(t+u)^\top] = \sum_{s=-\infty}^{\infty} \Lambda(s) \sum_{v=-\infty}^{\infty} \Gamma_{\mathbf{f}}(u+s-v)\Lambda(v)^\top + \Gamma_{\varepsilon}(u),$$

$$S_{\mathbf{X}}(\omega) = (2\pi)^{-1} \sum_{u=-\infty}^{\infty} \Gamma_{\mathbf{X}}(u)\exp(-i\omega u)$$
$$= \tilde{\Lambda}(\omega)S_{\mathbf{f}}(\omega)\tilde{\Lambda}(\omega)' + S_{\varepsilon}(\omega), \ -\pi \le \omega \le \pi. \quad (11.12)$$

In (11.12), $\tilde{\Lambda}(\omega) = \sum_{s=-\infty}^{\infty} \Lambda(s)\exp(-i\omega s)$ and the prime stands for transposition and conjugation. For statistical inference, it is important that identifiability conditions are imposed such that the representation in (11.12) is unique (up to scaling), see Geweke and Singleton (1981) for details. If the model is identified, the statistical parameter $\vartheta(\omega)$ (say) of interest for given frequency ω consists of all $d = 2pk + k^2 + p$ distinct parameters in $\tilde{\Lambda}(\omega)$, $S_{\mathbf{f}}(\omega)$ and $S_{\varepsilon}(\omega)$, where each of the (in general) complex elements in $\tilde{\Lambda}(\omega)$ and $S_{\mathbf{f}}(\omega)$ is represented by a pair of real components in $\vartheta(\omega)$, corresponding to its real part and its imaginary part. Hence, Hannan (1973) and Geweke and Singleton (1981) assumed the existence of B disjoint frequency bands $\Omega_1, \ldots, \Omega_B$, such that $S_{\mathbf{X}}$ is approximately constant within each of these bands. Letting $\omega^{(b)}$ denote the center of band Ω_b and $\vartheta_b = \vartheta(\omega^{(b)})$, where $1 \le b \le B$, the final model dimension is equal to Bd, which is certainly only feasible for small values of p and large values of T. Therefore, Hannan (1973) and Geweke and Singleton (1981) studied an asymptotic setting where $T \to \infty$.

Maximum likelihood estimators $\hat{\vartheta}_b$ for the parameters ϑ_b, $1 \le b \le B$, can be computed by adapting the algorithm by Jöreskog (1969). Based on central limit theorems for time series in the frequency domain, Geweke and Singleton (1981) showed that

$$\hat{\vartheta}_b \underset{asympt.}{\sim} \mathcal{N}_d(\vartheta_b, \hat{V}_b), \ 1 \le b \le B, \quad (11.13)$$

where $\hat{V}_b$ denotes the estimated covariance matrix of $\hat{\vartheta}_b$. The result in (11.13), in connection with the fact that the vectors $\hat{\vartheta}_b$, $1 \le b \le B$, are asymptotically jointly uncorrelated with each other, is very helpful for testing linear (point) hypotheses. Such hypotheses are of the form

$$H : C\vartheta = \xi \quad (11.14)$$

with a contrast matrix $C \in \mathbb{R}^{r \times Bd}$, $\xi \in \mathbb{R}^r$ and ϑ consisting of all elements of all the vectors ϑ_b. The integer r denotes the number of restrictions imposed on ϑ by H, where $r < Bd$ and C is assumed to have rank r. Notice that, in contrast to (4.7), H is regarded as one single hypothesis, namely, the intersection of the rows in the system of equations in (11.14). Geweke and Singleton (1981) proposed the usage of Wald statistics in this context. The Wald statistic for testing H is given by

$$W = (C\hat{\vartheta} - \xi)^\top (C\hat{V}C^\top)^{-1}(C\hat{\vartheta} - \xi), \tag{11.15}$$

where $\hat{\vartheta}$ is built in analogy to ϑ and $\hat{V}$ is a block matrix built up from the band-specific matrices $\hat{V}_b$, $1 \leq b \leq B$. It is well-known that W is asymptotically equivalent to the likelihood ratio statistic for testing H. In particular, W is asymptotically χ^2-distributed with r degrees of freedom under the null hypothesis H, see Sect. 12.4.2 in Lehmann and Romano (2005). Wald statistics have the practical advantage that they can be computed easily, avoiding restricted (by H) maximization of the likelihood function.

In the remainder of this section, we exemplify how two scientific questions of interest in the statistical analysis of dynamic factor models of the form (11.10) can be formalized as multiple test problems and addressed by multiple test procedures with vectors of Wald statistics as test statistics. To this end, we follow the derivations of Dickhaus (2012).

11.4.1 Which of the Specific Factors have a Non-trivial Autocorrelation Structure?

Addressing this question is of interest, because presence of many coloured noise components may hint at further hidden common factors and therefore, the solution to this problem can be utilized for the purpose of model diagnosis in the spirit of a residual analysis and, hence, for the choice of k. For one specific factor ε_i, $1 \leq i \leq p$, we consider the linear hypothesis $H_i : C_{\text{Dunnett}}\mathbf{s}_{\varepsilon_i} = 0$ of a flat spectrum. The contrast matrix C_{Dunnett} is the "multiple comparisons with a control" contrast matrix with $B - 1$ rows and B columns, where in each row j the first entry equals $+1$, the $(j + 1)$-th entry equals -1 and all other entries are equal to zero. The vector $\mathbf{s}_{\varepsilon_i} \in \mathbb{R}^B$ consists of the values of the spectral density matrix S_ε corresponding to the i-th noise component, evaluated at the B centers $(\omega^{(b)} : 1 \leq b \leq B)$ of the chosen frequency bands. Denoting the subvector of $\hat{\vartheta}$ that corresponds to $\mathbf{s}_{\varepsilon_i}$ by $\hat{\mathbf{s}}_{\varepsilon_i}$, the i-th Wald statistic is given by

$$W_i = (C_{\text{Dunnett}}\,\hat{\mathbf{s}}_{\varepsilon_i})^\top \left[C_{\text{Dunnett}}\hat{V}_{\varepsilon_i}C_{\text{Dunnett}}^\top\right]^{-1} (C_{\text{Dunnett}}\,\hat{\mathbf{s}}_{\varepsilon_i}),$$

where $\hat{V}_{\varepsilon_i} = \text{diag}(\hat{\sigma}^2_{\varepsilon_i}(\omega^{(b)}) : 1 \leq b \leq B)$. Then, under H_i, W_i asymptotically follows a χ^2-distribution with $B-1$ degrees of freedom if the corresponding limit matrix V_{ε_i} is assumed to be positive definite. Considering the vector $\mathbf{W} = (W_1, \ldots, W_p)^\top$ of all p Wald statistics corresponding to the p specific factors in the model and making use of the notation in Definition 4.6, we finally have $\mathbf{W} \underset{asympt.}{\sim} \chi^2(p, (B-1, \ldots, B-1)^\top, R)$ under the intersection H_0 of the p hypotheses $H_1, \ldots, H_p$, with some correlation matrix R. This distributional result allows for applying standard multiple tests that we have discussed in previous chapters.

11.4.2 Which of the Common Factors have a Lagged Influence on Which X_i?

In many applications, it is informative if certain factors have an instantaneous or a lagged effect. Here, we aim at addressing this question for all common factors simultaneously. As done by Geweke and Singleton (1981), we formalize the hypothesis that common factor j has a purely instantaneous effect on $\mathbf{X}_i$, $1 \leq j \leq k$, $1 \leq i \leq p$, in the spectral domain by

$$H_{ij} : |\tilde{\Lambda}_{ij}|^2 \text{ is constant across the } B \text{ frequency bands.}$$

In an analogous manner to the derivations in Sect. 11.4.1, the contrast matrix C_{Dunnett} can be used as the basis to construct a Wald statistic W_{ij}. The vector $\mathbf{W} = (W_{ij} : 1 \leq i \leq p, 1 \leq j \leq k)$ then asymptotically follows a multivariate chi-square distribution with $B-1$ degrees of freedom in each marginal under the corresponding null hypotheses and we can proceed as described in Sect. 11.4.1.

Many other problems of practical relevance can be formalized analogously by making use of linear contrasts and thus, the described framework applies to them, too. Furthermore, the hypotheses of interest may also refer to different subsets of $\{1, \ldots, B\}$. In such a case, the marginal degrees of freedom for the test statistics are not balanced, as considered in our general Definition 4.6.

Acknowledgments Parts of Sect. 11.2 have been the topic of Konstantin Schildknecht's diploma thesis and I thank him for discussing this section with me. Parts of Sect. 11.4 have profited from fruitful discussions with Markus Pauly and researchers from the Collaborative Research Center 649 "Economic Risk". Special thanks are due to Ruth Heller for personal communication regarding Benjamini and Heller (2007).

References

Benjamini Y, Heller R (2007) False discovery rates for spatial signals. J Am Stat Assoc 102(480):1272–1281. doi:10.1198/016214507000000941

Benjamini Y, Heller R (2008) Screening for partial conjunction hypotheses. Biometrics 64(4): 1215–1222. doi:10.1111/j.1541-0420.2007.00984.x

Benjamini Y, Hochberg Y (1997) Multiple hypotheses testing with weights. Scand J Stat 24(3): 407–418. doi:10.1111/1467-9469.00072

Benjamini Y, Krieger AM, Yekutieli D (2006) Adaptive linear step-up procedures that control the false discovery rate. Biometrika 93(3):491–507

Blanchard G, Delattre S, Roquain E (2013) Testing over a continuum of null hypotheses. Bernoulli forthcoming.

Bogomolov M (2011) Testing of Several Families of Hypotheses. PhD thesis, Tel-Aviv University

van Bömmel A, Song S, Majer P, Mohr PNC, Heekeren HR, Härdle WK (2013) Risk Patterns and Correlated Brain Activities. Multidimensional statistical analysis of fMRI data in economic decision making study, Psychometrika forthcoming

Dickhaus T (2012) Simultaneous Statistical Inference in Dynamic Factor Models. SFB 649 Discussion Paper 2012–033, Sonderforschungsbereich 649, Humboldt Universität zu Berlin, Germany, available at http://sfb649.wiwi.hu-berlin.de/papers/pdf/SFB649DP2012-033.pdf
Friston KJ, Penny WD, Glaser DE (2005) Conjunction revisited. NeuroImage 25(3):661–667
Genovese CR, Lazar NA, Nichols T (2002) Thresholding of statistical maps in functional neuroimaging using the false discovery rate. NeuroImage 15(4):870–878
Geweke JF, Singleton KJ (1981) Maximum likelihood "confirmatory" factor analysis of economic time series. Int Econ Rev 22:37–54. doi:10.2307/2526134
Hallin M, Lippi M (2013) Factor models in high-dimensional time series - A time-domain approach. Stoch Process Appl 123(7):2678–2695
Hannan E (1973) Central limit theorems for time series regression. Z Wahrscheinlichkeitstheor Verw Geb 26:157–170. doi:10.1007/BF00533484
Heller R, Stanley D, Yekutieli D, Rubin N, Benjamini Y (2006) Cluster-based analysis of FMRI data. NeuroImage 33(2):599–608
Hu JX, Zhao H, Zhou HH (2010) False Discovery Rate Control With Groups. J Am Stat Assoc 105(491):1215–1227
Jöreskog KG (1969) A general approach to confirmatory maximum likelihood factor analysis. Psychometrika 34(2):183–202
Lazar NA (2008) The statistical analysis of functional MRI data. Statistics for biology and health. Springer, New York
Lehmann EL, Romano JP (2005) Testing statistical hypotheses. Springer texts in statistics, 3rd edn. Springer, New York
Pacifico MP, Genovese C, Verdinelli I, Wasserman L (2004) False discovery control for random fields. J Am Stat Assoc 99(468):1002–1014. doi:10.1198/0162145000001655
Park BU, Mammen E, Härdle W, Borak S (2009) Time series modelling with semiparametric factor dynamics. J Am Stat Assoc 104(485):284–298
Peña D, Box GEP (1987) Identifying a simplifying structure in time series. J Am Stat Assoc 82: 836–843. doi:10.2307/2288794
Poldrack RA (2007) Region of interest analysis for fMRI. Soc Cogn Affect Neurosci 2(1):67–70
Taylor JE, Worsley KJ (2007) Detecting sparse signals in random fields, with an application to brain mapping. J Am Stat Assoc 102(479):913–928. doi:10.1198/016214507000000815
Worsley KJ (2003) Detecting activation in fMRI data. Stat Methods Med Res 12(5):401–418
Worsley KJ, Andermann M, Koulis T, MacDonald D, Evans AC (1999) Detecting changes in nonisotropic images. Hum Brain Mapp 8(2–3):98–101
Worsley KJ, Liao CH, Aston J, Petre V, Duncan GH, Morales F, Evans AC (2002) A general statistical analysis for fMRI data. NeuroImage 15(1):1–15

Part III
Further Applications in the Life Sciences

Chapter 12
Further Life Science Applications

Abstract In this concluding chapter, we exemplarily discuss binary classification and multiple test problems in two specific areas of the life sciences, namely, brain-computer interfacing (BCI) and gel electrophoresis-based proteome analysis. In the BCI context, we demonstrate how multiple testing-based approaches to binary classification that we have described in Chap. 6 can be utilized for data analysis. In particular, estimation methods for multivariate stationary densities of autocorrelated time series vectors are employed. In the context of proteome analysis, the application of multiple tests is demonstrated by means of analyzing a real-life dataset from diabetes research. Finally, an outlook to further applications in the respective fields is provided.

This concluding chapter is concerned with two fields of life science for which the application of simultaneous statistical inference methods that we have described in Part I is not so well-established yet. The models and procedures considered in Sects. 12.1 and 12.2 below are therefore not meant to describe the state-of-the-art in the respective field, but the chapter attempts to provide an outlook on potential application fields of simultaneous statistical inference which may attract further attention in the future.

12.1 Brain-Computer Interfacing

Brain-computer interfaces (BCIs) are systems that convert brain activity in real-time into control signals for a computer application. This can allow, for instance, paralyzed patients who are deprived of other means of communication to interact with the external world via BCI-controlled rehabilitative tools. General introductions to BCIs and their applications have been provided by Dornhege et al. (2007), Graimann et al. (2010) and Wolpaw and Wolpaw (2012). Recently, also nonmedical applications of BCI technologies are being explored, see Blankertz et al. (2010b).

T. Dickhaus, *Simultaneous Statistical Inference*,
DOI: 10.1007/978-3-642-45182-9_12,

The neural activity of the central nervous system can be acquired in different ways. Most BCIs employ the noninvasive electroencephalogram (EEG) or implanted electrode arrays. There is also a large variety of different control strategies, which allow the user to generate specific brain signals that have relatively good detection rates by the BCI. In the implementation of a BCI system one has to use an algorithm which extracts features that correspond to the chosen control strategy (see, e.g., Blankertz et al. 2008, 2011). The choice of suitable control strategies and corresponding feature extraction algorithms typically profits from neurophysiological background knowledge. The Berlin brain-computer interface group (see Blankertz et al. 2010a) employs statistical (machine) learning methods for feature extraction and processing. This approach is meant to avoid extensive training on the users' side and to transfer the largest amount of workload to the machines ("let the machines learn"). A general introduction to machine learning methods for EEG analyses has been provided by Lemm et al. (2011).

Here, let us focus our attention on one prominent classification problem (cf. Chap. 6) in the BCI field. It is known for a long time that the imagination of hand movements corresponds to specific changes of the brain signals that can be detected at a macroscopic level, i.e., with noninvasive EEG recordings from the scalp. In particular, the oscillations which can be observed during idle state in the brain area corresponding to the respective (left or right) hand are attenuated during motor imagery—an effect called event-related desynchronization (ERD) of sensorimotor EEG rhythms, see Pfurtscheller and da Silva (1999). Since the two hand areas are spatially well separated in the motor cortex of the brain, the ERD effects can, in principle, be distinguished to originate either from the left or from the right hand. Therefore, one type of BCI can be realized in the following way: the user switches voluntarily between motor imagery of the left hand and the right hand and thereby continuously transmits a binary control signal. Hence, even if actual motor functions of the user are impaired, ERD-based BCIs continue to work if s/he can at least imagine to move his/her hands. This can be used, e.g., for one-dimensional cursor control (see Blankertz et al. 2007) or, with an intelligent design, for BCI-assisted typewriting (see Williamson et al. 2009).

Dickhaus et al. (2013) demonstrated how EEG features from a BCI experiment relying on ERD of sensorimotor rhythms can be classified by making use of multiple testing-based Algorithms 6.1 and 6.2 that we have described in Sect. 6.2. The observational units of this classification problem are those segments of the (band-pass filtered) multichannel EEG data during which motor imagery was performed. Each of the resulting so-called "single trials" results in a matrix $\Xi_i \in \mathbb{R}^{C \times T}$ (say), where C denotes the number of recorded EEG channels (which is 86 in our example) and T the number of sampled time points within each trial (which is 400 in our example: four seconds at a sampling rate of 100 Hz). The entries of Ξ_i are measurements of the electrical activity in the corresponding channel at the corresponding time point. The index i refers to the number of the single trial. As described around Algorithms 6.1 and 6.2, the classification parameters f_0, f_1 and $c(w)$ have been learned utilizing a training (calibration) dataset. From these calibration data, spatial filters $w_j \in \mathbb{R}^C$ for $1 \leq j \leq k$ that are optimized for the discrimination of left and right hand trials were determined by Common Spatial Pattern (CSP) analysis (see Fukunaga 1990). The number k of spatial filters that are employed is data-dependent and varies in our approach, as described by Blankertz et al. (2008), between 2 and 6. This number k corresponds to the data space dimensionality considered in our general classification setup, cf. Definition 6.1. Features are then calculated as the logarithmic bandpower in the spatially filtered channels. Thus, we have k features $x_{i,j} = \log(w_j^T \Xi_i \Xi_i^T w_j)$, $j = 1, \ldots, k$, in trial i, forming a feature vector $\mathbf{x}_i \in \mathbb{R}^k$. Assuming that we have conducted m_{train} calibration trials for one BCI user, this results in a training dataset $(\mathbf{x}_1, y_1), \ldots, (\mathbf{x}_{m_{\text{train}}}, y_{m_{\text{train}}})$, where $y_i \in \{0, 1\}$ is the class label of trial i with *left* and *right* hand motor imagery being coded by 0 and 1, respectively.

Certainly, the aforementioned EEG features $\mathbf{x}_i$ are autocorrelated. Hence, as outlined in the discussion below Remark 6.2, a generalization of the classification model defined in Definition 6.1 is necessary in order to account at least for weak autocorrelations of features. Dickhaus et al. (2013) defined the following weakly dependent mixture model for classification.

Definition 12.1 (Dickhaus et al. 2013). Assume that m_{train} training trials for a given classification problem have been performed, resulting in a training dataset $(\mathbf{x}_1, y_1), \ldots, (\mathbf{x}_{m_{\text{train}}}, y_{m_{\text{train}}})$. Furthermore, assume that m test data points $\mathbf{x}_{m_{\text{train}}+1}, \ldots, \mathbf{x}_{m_{\text{train}}+m}$ with unknown labels have to be classified, and let $M = m_{\text{train}} + m$. Then, the multivariate distribution of the data tuples $\{(\mathbf{X}_i, Y_i)\}_{i=1}^M$ on $(\mathbb{R}^k \times \{0, 1\})^M$ is called a weakly dependent mixture model if the conditions (12.1)–(12.3) are fulfilled.

$$\text{Defining } \gamma_M = m_{\text{train}}/M, \text{ it holds } 0 < \liminf_{M\to\infty} \gamma_M \le \limsup_{M\to\infty} \gamma_M < 1. \tag{12.1}$$

Independently of the values in the sequence $\{\gamma_M\}_M$, it holds

$$M^{-1} \sum_{i=1}^{M} (1 - Y_i) \to \pi_0 = 1 - \pi_1 \in (0, 1) \text{ almost surely.} \tag{12.2}$$

There exist continuous cdfs $F_0 \neq F_1$ on $\mathbb{R}^k$, not depending on the values in the sequence $\{\gamma_M\}_M$, with the property that for all $\mathbf{x} \in \mathbb{R}^k$ it holds

$$M_0^{-1} \sum_{i=1}^{M} (1 - Y_i)\, \mathbf{1}\{\mathbf{X}_i \le \mathbf{x}\} \to F_0(\mathbf{x}) \text{ and } M_1^{-1} \sum_{i=1}^{M} Y_i\, \mathbf{1}\{\mathbf{X}_i \le \mathbf{x}\} \to F_1(\mathbf{x}), \tag{12.3}$$

almost surely, with M_0 and M_1 denoting the total number of trials with $y_i = 0$ and $y_i = 1$, respectively.

The distributional assumptions in Definition 12.1 imply that the Y_i asymptotically follow a Bernoulli(π_1)-distribution and the $\mathbf{X}_i$ are asymptotically distributed according to the mixture cdf $(1 - Y_i)F_0 + Y_i F_1$. Under these assumptions, at least for large training data sets, the asymptotic pdfs f_0 and f_1 can consistently be estimated from the training data where labeling is known and separate estimation of the densities corresponding to $y_i = 0$ and $y_i = 1$, respectively, is possible. Notice that Definition 12.1 is only concerned with the dependency structure between distinct random vectors, while the "inner-vector" dependency structure, i.e., the dependence between the components of a particular $\mathbf{X}_i$, is not considered. The latter dependency is implicitly addressed in density (ratio) estimation, cf. Remark 6.2. In our case, empirically sphered data were used as input of a kernel density estimator.

Under the assumptions of Definition 12.1, Dickhaus et al. (2013) applied Algorithms 6.1 and 6.2 to a BCI dataset described by Dickhaus et al. (2009). The two multiple testing-based classification procedures exhibited classification performances which were comparable to those of regularized linear discriminant analysis, a method known to perform well on log-transformed EEG bandpower features, see Blankertz et al. (2008, 2011). This is a remarkable result, because the classification approach making use of Algorithms 6.1 and 6.2 avoids relying on such prior knowledge, since it is fully nonparametric. Therefore, it may also be applied to data from new experimental paradigms for which no laboratory experience is existing yet.

Remark 12.1. Actual multiple testing problems in the context of EEG analyses and brain-computer interfaces have been discussed by Hemmelmann et al. (2005), Hemmelmann et al. (2008), Singh and Phillips (2010), Billinger et al. (2012) and Milekovic et al. (2012), among others.

12.2 Gel Electrophoresis-Based Proteome Analysis

Two-dimensional electrophoresis (2-DE) is a technique to measure the abundance of several types of proteins in one experiment. Molecules are moved through a gel matrix by applying an electric field. Depending on their mass and their isoelectric point, their movement speed and their final position differs. This can be exploited in order to quantify the amount of specific proteins in the probe. The resulting measurements originate from image processing and are referred to as spot intensities, where typically each spot on the gel corresponds to one protein. For some technical details see, for instance, Bajla et al. (2001). Similarly to the analysis of differential gene expression (see Chap. 10), applying this technique to material from different populations or different experimental conditions leads to a family of tests for differences between groups with respect to many endpoints, where every endpoint is given by one spot (protein). Diz et al. (2011) emphasized the importance of applying multiple testing methods for the statistical analysis.

Example 12.1. Dickhaus (2008) describes a proteomics experiment in which 1330 protein spots from two groups A and B were detected and matched by a spot detection software. The protein material consisted of pooled tissue from two different mice stems under investigation in a diabetes-specific context. Aim of the statistical

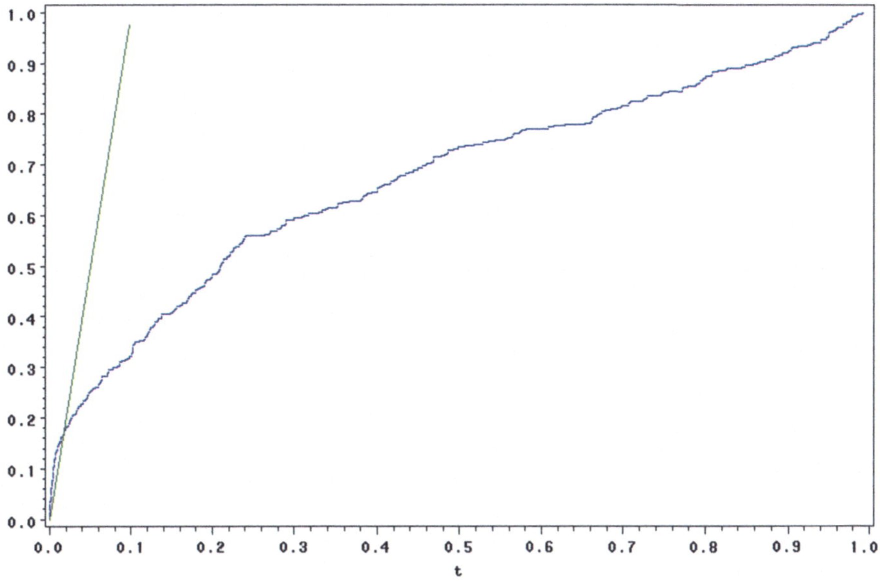

Fig. 12.1 Simes' rejection line for $\alpha = 0.1$ and ecdf. of 393 marginal p-values from Example 12.1

analysis was to detect differences with respect to spot intensities in the two groups. Group A was processed on four independent gels and group B was processed on three independent gels (the fourth one for group B was defective). During data cleaning and preparation, only spots with a minimal measurement number of three per group were retained. Furthermore, intensities below 0.5 were excluded because of lacking reliability and relevance. As in gene expression data, often a log-normal distribution for the intensity ratios is assumed. Therefore, the remaining intensities were transformed by applying the natural logarithm. After these steps, $m = 393$ spots remained. Diagnostic plots justified the normal distribution assumption for these remaining log-intensities and therefore, two-sided two-sample t-tests for the logarithmic intensity differences per spot were carried out (cf. part (a) of Definition 10.1). This resulted in $m = 393$ marginal p-values. Figure 12.1 displays Simes' rejection line for $\alpha = 0.1$ (cf. the discussion around Lemma 5.6) and the ecdf. of the obtained $m = 393$ p-values. The crossing point of these two objects determines the decision rule of the linear step-up test φ^{LSU} from Definition 5.6. In this example, 64 hypotheses were rejected at FDR level 0.1. The proportion $R_m/m = 0.163$ of rejected hypotheses is given by the ordinate of the crossing point.

Further references for multiple testing theory and applications in gel electrophoresis-based proteome analysis comprise Morris et al. (2011), Morris (2013), Langley et al. (2013), Corzett et al. (2006), Corzett et al. (2010), Rabilloud (2012) and references therein, among others.

Remark 12.2. Often, detection of differentially expressed proteins by analyzing spots in 2-DE is performed in connection with a following mass spectrometry analysis of the detected spots. Analyzing peaks in protein mass spectra and comparing two or more of such spectra constitute two further applications of multiple testing in proteome analysis which are, however, rather different from what we have described in the present section. One multiple testing-related software package for the analysis of protein mass spectra is `MALDIquant`, described by Gibb and Strimmer (2012).

Acknowledgments Parts of Sect. 12.1 originated from joint work with the Berlin brain-computer interface group, in particular with Benjamin Blankertz and Frank C. Meinecke. Data for Example 12.1 have been generated in the Institute for Clinical Biochemistry and Pathobiochemistry of the German Diabetes Center Düsseldorf.

References

Bajla I, Holländer I, Burg K (2001) Improvement of electrophoretic gel image analysis. Measur Sci Rev 1(1):5–10

Billinger M, Brunner C, Scherer R, Holzinger A, Müller-Putz GR (2012) Towards a framework based on single trial connectivity for enhancing knowledge discovery in BCI. In: Huang R, Ghorbani A, Pasi G, Yamaguchi T, Yen N, Jin B (eds) Active media technology. Lecture notes in computer science, vol 7669. Springer, New York, pp 658–667

Blankertz B, Dornhege G, Krauledat M, Müller KR, Curio G (2007) The non-invasive Berlin brain-computer interface: fast acquisition of effective performance in untrained subjects. NeuroImage 37(2):539–550

Blankertz B, Tomioka R, Lemm S, Kawanabe M, Müller KR (2008) Optimizing spatial filters for robust EEG single-trial analysis. IEEE Sig Process Mag 25(1):41–56. doi:10.1109/MSP.2008.4408441

Blankertz B, Tangermann M, Vidaurre C, Dickhaus T, Sannelli C, Popescu F, Fazli S, Danóczy M, Curio G, Müller KR (2010a) Detecting mental states by machine learning techniques: the Berlin brain-computer interface. In: Graimann B, Allison B, Pfurtscheller G (eds) Brain-computer interfaces: revolutionizing human-computer interaction. Springer, Berlin, pp 113–135

Blankertz B, Tangermann M, Vidaurre C, Fazli S, Sannelli C, Haufe S, Maeder C, Ramsey LE, Sturm I, Curio G, Müller KR (2010b) The Berlin brain-computer interface: non-medical uses of BCI technology. Front Neurosci 4:198. doi:10.3389/fnins.2010.00198

Blankertz B, Lemm S, Treder MS, Haufe S, Müller KR (2011) Single-trial analysis and classification of ERP components—a tutorial. NeuroImage 56:814–825. doi:10.1016/j.neuroimage.2010.06.048

Corzett TH, Fodor IK, Choi MW, Walsworth VL, Chromy BA, Turteltaub KW, McCutchen-Maloney SL (2006) Statistical analysis of the experimental variation in the proteomic characterization of human plasma by two-dimensional difference gel electrophoresis. J Proteome Res 5(10):2611–2619

Corzett TH, Fodor IK, Choi MW, Walsworth VL, Turteltaub KW, McCutchen-Maloney SL, Chromy BA (2010) Statistical analysis of variation in the human plasma proteome. J Biomed Biotechnol 258:494

Dickhaus T (2008) False discovery rate and asymptotics. PhD thesis, Heinrich-Heine-Universität Düsseldorf.

Dickhaus T, Sannelli C, Müller KR, Curio G, Blankertz B (2009) Predicting BCI performance to study BCI illiteracy. BMC Neurosci 10(Suppl 1):P84. doi:10.1186/1471-2202-10-S1-P84

Dickhaus T, Blankertz B, Meinecke FC (2013) Binary classification with pFDR-pFNR losses. Biom J 55(3):463–477. doi:10.1002/bimj.201200054

Diz AP, Carvajal-Rodriguez A, Skibinski DO (2011) Multiple hypothesis testing in proteomics: a strategy for experimental work. Mol Cell Proteomics 10(3):M110.004374

Dornhege G, del R Millán J, Hinterberger T, McFarland D, Müller KR (eds) (2007) Toward Brain-computer interfacing. MIT Press, Cambridge

Fukunaga K (1990) Introduction to statistical pattern recognition, 2nd edn. Academic Press, Boston

Gibb S, Strimmer K (2012) MALDIquant: a versatile R package for the analysis of mass spectrometry data. Bioinformatics 28(17):2270–2271

Graimann B, Allison B, Pfurtscheller G, (eds), (2010) Brain-computer interfaces: revolutionizing human-computer interaction. Springer, Berlin. ISBN 13: 978-3-642-02090-2

Hemmelmann C, Horn M, Susse T, Vollandt R, Weiss S (2005) New concepts of multiple tests and their use for evaluating high-dimensional EEG data. J Neurosci Methods 142(2):209–217

Hemmelmann C, Ziegler A, Guiard V, Weiss S, Walther M, Vollandt R (2008) Multiple test procedures using an upper bound of the number of true hypotheses and their use for evaluating high-dimensional EEG data. J Neurosci Methods 170(1):158–164

Langley SR, Dwyer J, Drozdov I, Yin X, Mayr M (2013) Proteomics: from single molecules to biological pathways. Cardiovasc Res 97(4):612–622

Lemm S, Blankertz B, Dickhaus T, Müller KR (2011) Introduction to machine learning for brain imaging. NeuroImage 56:387–399. doi:10.1016/j.neuroimage.2010.11.004

Milekovic T, Fischer J, Pistohl T, Ruescher J, Schulze-Bonhage A, Aertsen A, Rickert J, Ball T, Mehring C (2012), An online brain-machine interface using decoding of movement direction from the human electrocorticogram. J Neural Eng 9(4):046003

Morris JS (2012) Statistical methods for proteomic biomarker discovery based on feature extraction or functional modeling approaches. Stat Interface 5(1):117–135

Morris JS, Baladandayuthapani V, Herrick RC, Sanna P, Gutstein H (2011) Automated analysis of quantitative image data using isomorphic functional mixed models, with application to proteomics data. Ann Appl Stat 5(2A):894–923. doi:10.1214/10-AOAS407

Pfurtscheller G, da Silva FHL (1999) Event-related EEG/MEG synchronization and desynchronization: basic principles. Clin Neurophysiol 110(11):1842–1857

Rabilloud T (2012) The whereabouts of 2D gels in quantitative proteomics. Methods Mol Biol 893:25–35

Singh AK, Phillips S (2010) Hierarchical control of false discovery rate for phase locking measures of EEG synchrony. Neuroimage 50(1):40–47

Williamson J, Murray-Smith R, Blankertz B, Krauledat M, Müller KR (2009) Designing for uncertain, asymmetric control: interaction design for brain-computer interfaces. Int J Hum Comput Stud 67(10):827–841

Wolpaw JR, Wolpaw EW (eds) (2012) Brain-computer interfaces : principles and practice. Oxford University Press, ISBN-13: 978–0195388855

Index

T. Dickhaus, *Simultaneous Statistical Inference*,
DOI: 10.1007/978-3-642-45182-9,

The manufacturer's authorised representative in the EU is Springer Nature Customer Service Centre GmbH, Europaplatz 3, 69115 Heidelberg, Germany. If you have any concerns regarding our products, please contact ProductSafety@springernature.com

Printed and bound by CPI Group (UK) Ltd, Croydon, CR0 4YY
01/10/2026
02225973-0001